OHIO BIOLOGICAL SURVEY
MISCELLANEOUS CONTRIBUTIONS NO. 3

POPULATION TRENDS OF BREEDING BIRDS IN OHIO

by

Susan L. Earnst
and
Brad A. Andres

Ohio Cooperative Fish and Wildlife Research Unit
National Biological Service

Published By
OHIO BIOLOGICAL SURVEY
COLLEGE OF BIOLOGICAL SCIENCES
THE OHIO STATE UNIVERSITY

In Cooperation With
OHIO DIVISION OF WILDLIFE and NATIONAL BIOLOGICAL SERVICE

COLUMBUS, OHIO 43212-1192   1996

OHIO BIOLOGICAL SURVEY

Miscellaneous Contributions Series ISSN: 1074-9233
Miscellaneous Contributions Number 3 ISBN: 0-86727-121-5
Library of Congress Number: 96-068822

EDITOR

Veda M. Cafazzo

CITATION

Earnst, Susan L. and Brad A. Andres. 1996. Population Trends of Breeding Birds in Ohio. Ohio Biol. Surv. Misc. Cont. No. 3 vii + 125 p.

COVER ILLUSTRATION—Carolina Wren, *Thryothorus ludovicianus* (Latham). Original photograph by Al Staffen, staff photographer, Ohio Department of Natural Resources.

Color Production:	Cardinal Imaging, Columbus, OH
Printing:	The Ohio State University Printing Services
Binding:	H. & H. Bookbinders, Indianapolis, IN

Ohio Biological Survey
College of Biological Sciences
The Ohio State Univesity
Columbus, Ohio 43212-1192
1996

6-96—2M

ABSTRACT

   Breeding Bird Survey data from 1966 through 1994 were used to estimate the population trends of breeding birds in Ohio. An interpretation and graphical presentation of the trend and a map of the relative abundance of individuals per route are provided for each species. Overall, 21% of Ohio's breeding birds decreased significantly and 45% increased significantly. Neotropical migrants do not appear to have declined proportionately more than temperate migrants or residents; 11 neotropical migrants (23%) and 10 temperate migrants or residents (20%) declined significantly. Grassland breeding species exhibited proportionately more declines than did species breeding in other habitats. Ten of the 14 (71%) grassland species declined significantly and the ratio of significantly declining to increasing species was 10:1. Mature woodlands had the lowest ratio of significantly declining to increasing species (1:6).

ACKNOWLEDGEMENTS

The Breeding Bird Survey would not be possible without the help of the dedicated volunteer observers who survey routes each spring. Their efforts allow us to detect, and find methods to reverse, declines in bird populations. The following individuals surveyed one or more routes during 1966-1994: G.H. Ash, R.E. Ashton, C.S. Banks, W.H. Barnes, L.M. Barnhart, J.R. Bart, H.T. Bartlett, M. Bell, G.E. Bernhardt, D.J. Borror, R.A. Bradley, K.A. Bricker, D.A. Brinkman, A.S. Brooks, R.W. Bryant, J. Buecking, E.H. Burtt, J.A. Butts, M.L. Carpenter, D.S. Case, C. Cathers, C.F. Clark, G. Cramer, R.J. Crofts, A.W. Cusick, P.M. Daniel, D. Dister, G. Doig, R.A. Dolbeer, R.W. Donohoe, C.L. Dusthimer, R.A. Eberhardt, R. Frankhouse, K.M. Gale, L. Gara, L.A. Garling, R.D. Glick, E.E. Good, H.L. Gratz, M.E. Gustafson, E.F. Hardesty, R.N. Harlan, J.G. Hawkins, J.A. Hickman, J.H. Hill, H.W. Hintz, L.T. Holtzman, H. Honschop, J.L. Ingold, R.F. Jezerinac, R.C. Jones, I. Kassoy, J.M. Keener, T.R. Kemp, J. Kiefer, W.A. Klamm, I.H. Klein, J.S. Knoblaugh, P.E. Knoop, Jr., L.J. Laux, Jr., E.N. Leibold, R.X. Little, E.W. Martin, C. Mathena, T.O. Matson, J.S. McCormac, C.R. McCullough, R.S. McCutchen, S.M. McKee, B.B. McLean, B.J. Miller, D. Minney, M.A. Moeckel, S.F. Moeckel, N.J. Moore, B. Murphy, P.B. Murphy, J.A. Myers, A.V. Newell, K.A. Noblet, M.J. Norrocky, L.P. Orr, D. Overacker, J.W. Parrish, Jr., B.G. Peterjohn, T. Peterson, R.S. Phillips, W. Randle, R.D. Reed, C.R. Reese, D.L. Rice, J.C. Ritzenthaler, L.O. Rosche, P.J. Rowe, E.R. Schlabach, H.C. Seibert, B.H. Sellner, M.E. Shrader, J.P. Sipe, S. Sjodahl, J.A. Smallwood, G.H. Smith, J.E. Stahl, J.J. Stenger, A.R. Stickley, Jr., R.J. Stoll, J.J. Stophlet, W.D. Stull, W.L. Sturm, D. Styer, J.R. Talkington, M. Tawse, T.D. Taylor, B.G. Thobaben, Jr., R. Thoma, D.E. Todt, K.R. Troutman, L.F. Vancamp, D.W. Waller, B.J. Waterbury, J.T. Watts, P. Whan, C.B. Wheeler, A.J. Wiseman, D. Wood, M.M. Wyatt.

Many thanks to Denis Case who provided the original idea and impetus for this project and who painstakingly edited several drafts of this publication. We thank Jonathan Bart for advice throughout the project and editorial comments on an earlier version of the manuscript, Bruce G. Peterjohn and the Breeding Bird Survey Office, National Biological Service, for providing the Ohio BBS data and the continental and regional population trends, and Veda M. Cafazzo and Brian Armitage for editorial assistance and support throughout the project. This project was funded by the Ohio Division of Wildlife and the Wildlife Diversity Tax Checkoff.

Susan L. Earnst, Assistant Leader and Assistant Professor, Ohio Cooperative Fish and Wildlife Research Unit, National Biological Service, and Department of Zoology, The Ohio State University, 1735 Neil Avenue, Columbus, OH 43210

Brad A. Andres, Wildlife Biologist, Migratory Bird Management, U.S. Fish and Wildlife Service, 1011 E. Tudor Road, Anchorage, AK 99503

CONTENTS

PASSERIFORMES, continued.

INTRODUCTION

Recent declines in songbird populations have generated concern in the lay and scientific community. Although the decline of neotropical migrants has received the most attention thus far, several studies have emphasized that patterns of population trends vary by geographic region (see below). The present study was undertaken to determine the population trends of Ohio's breeding birds and thus assist managers in identifying those species and habitats in most immediate need of conservation.

The proportionately greater decline of neotropical migrants compared to temperate migrants or residents is most evident in forest-breeding species during the last 15 years or so, especially in the eastern U.S. (Robbins et al., 1989; Sauer and Droege, 1992; and Peterjohn and Sauer, 1994). However, there is evidence that non-forest breeding species should be of more concern in some areas. For example, scrub-breeding species appear to be declining as much or more than forest-breeding species in the central United States (Sauer and Droege, 1992), southeast and southcentral United States (James et al., 1992), and areas of New England (Witham and Hunter, 1992). Even more striking, grassland breeding species have incurred more drastic and more widespread declines than any other group (Knopf, 1994). Similarly, there is growing evidence that temperate migrants are declining in equal or greater proportion to neotropical migrants in some areas and habitats (Hagan et al., 1992; Witham and Hunter, 1992; and Bohning-Gaese et al., 1993).

In Ohio, preliminary evidence based on North American Breeding Bird Survey (BBS) data (Andres, 1989), the same data used in the studies above, suggested that patterns of population trends were different than in the eastern U.S. as a whole, and that several species of interest were not well-monitored by current surveying techniques. Thus, the objectives of the present study are to (1) use BBS data to determine long-term population trends of Ohio's breeding birds, and (2) identify species for which BBS data are inadequate to estimate the population trend and thus for which alternative monitoring systems should be used.

METHODS

OVERVIEW OF THE NORTH AMERICAN BREEDING BIRD SURVEY

The North American Breeding Bird Survey was designed to monitor populations and assess the relative abundance of breeding birds in the United States and Southern Canada. The BBS is a joint project of the U.S. Fish and Wildlife Service and the Canadian Wildlife Service and consists of over 3000 roadside routes throughout the U.S. and Southern Canada. The number of routes varies among geographic regions (i.e., the Midwest vs. Eastern U.S.) depending on the availability of qualified, volunteer observers and accessibility via secondary roads. In the Eastern U.S., most routes were established in 1966, and most in western states were established by 1968. Routes were placed on secondary roads randomly selected from each 1° latitude-longitude block. Each route is 24.5 miles long and consists of 50 stops located at 0.5 mile intervals. At each stop, the observer records number of individuals of each species heard or seen within a 0.25-mile radius during a 3-minute period. Routes are surveyed each year at the beginning of the breeding season (early June in Ohio), beginning one-half hour before sunrise. A single observer is encouraged to survey the same route for as many consecutive years as possible. The BBS protocol is reviewed more thoroughly in Robbins et al. (1986).

Population trends estimated from BBS data can be affected by factors which influence local but not regional abundance, such as changes in roadside habitat (Bart et al., 1995) and factors which affect the detectability of birds present, such as weather or differences in observer ability between years. Change in observer ability is the most serious potential bias and is discussed further in the Appendix and by Bart and Schoultz (1984), Robbins et al. (1986), Sauer and Bortner (1990), and Sauer et al. (1994). Despite these constraints, BBS data provide good estimates of long-term population trends of most species, especially those that are relatively abundant and widespread in the geographic area of interest.

THE BREEDING BIRD SURVEY IN OHIO

In Ohio, data collection on 45 primary BBS routes was initiated in 1966. Seven replacement routes have been established since 1980 to replace routes along roads that had become impassable, unsafe, or so noisy that the detectability of singing birds was substantially reduced. Prior to analysis, we combined data from each replacement route and its respective primary route. The analysis presented here is based solely on the original

Fig.1. Locations of the 45 Breeding Bird Survey routes and the physiographic regions comprising Western Ohio (I - Great Lakes, II - Till Plain, III - Lexington Plains) and Eastern Ohio (IV - Allegheny Plateau, V - Ohio Hills).

45 routes and their replacement routes. The analysis does not include any recently established primary routes nor the recently established nonrandom routes that were added to increase the coverage of particular habitats.

Of the 45 primary BBS routes, 21 are located in the predominately forested area of Eastern Ohio and 24 in the grasslands and agricultural lands of Western Ohio (Figure 1). We delineate Western Ohio as the Great Lakes, Till, and Lexington Plains; and we delineate Eastern Ohio as the Allegheny Plateau and the Ohio Hills (physiographic strata defined by Robbins et al., 1986). Abundance and population trends were compared between Eastern and Western Ohio because the two regions differ strikingly in habitat composition and sample sizes were usually adequate; comparisons between Northern and Southern Ohio and other comparisons were made only when warranted. The boundary between Northern and Southern Ohio was defined as the northern border of the following counties, listed from west to east: Darke, Shelby, Logan, Union, Delaware, Licking, Coshocton, Guernsey, and Belmont.

Species that were recorded on < 14 routes were not included in the Species Accounts because we considered there to be too little data to reliably estimate a population trend (the 14-route guideline is also recommended by the BBS Office; e.g., Peterjohn et al., 1996). In addition, some species recorded on $\geq$ 14 routes, but recorded in few years or in low numbers, were also monitored inadequately (see **ADEQUACY OF BBS FOR DETECTING POPULATION TRENDS OF OHIO'S BREEDING BIRDS**), suggesting caution in interpreting their estimated population trends.

MIGRATORY STATUS

Each species was assigned to one of 4 migratory categories based on the most common wintering grounds of Ohio breeders. **Permanent residents** were defined as those species in which most individuals breeding in Ohio also winter in Ohio; **temperate migrants** as those that winter in the southern U.S.; **central neotropical migrants** in Mexico, Central America, and the Caribbean; and **southern neotropical migrants** in South America. Information on winter ranges was obtained from Rappole et al. (1983); the National Geographic Society (1987) field guide; Peterjohn (1989); and Peterjohn and Rice (1991).

Central and **southern neotropical migrants**, as defined here, include the Type A migrants of Partners in Flight, i.e., those that winter primarily south of the U.S., and temperate migrants include Type B migrants i.e., those that winter extensively in North America, but have some populations that winter south of the U.S. (Gauthreaux, 1991).

BREEDING HABITAT

Six primary categories of breeding habitat were delineated: **mature woodlands**, comprised of sawtimber and older trees, **young woodlands** comprised of poletimber trees, **scrub** defined as early successional seedling, brushy, and sapling stages, **grasslands**, **wetlands**, and **urban/residential**. Griffith et al. (1993) define sawtimber trees as $\geq$ 9 or $\geq$ 11 inches diameter-at-breast-height (dbh) for softwoods and hardwoods, respectively, and define poletimber trees as $\geq$ 5 inches dbh but less than sawtimber dbh. We added additional modifiers to our 6 habitat categories to indicate preferences for lowland forests, forest interiors, riparian corridors, edges, mixtures of habitats, etc. For the analysis of population trends by habitat, young woodlands and scrub were combined, and a **nonspecific woodland** category was included for those species which occur in both mature and young/scrub woodlands (i.e., were not assignable to one vs. the other) and for those which prefer an interspersed woodland-grassland mosaic (e.g., Red-tailed Hawks, Orchard Orioles). A few species could arguable have been placed in either of two habitat categories (e.g., Chipping Sparrows in residential or nonspecific woods); overall conclusions concerning population trends among habitats were not affected by transferring these few species among categories. Information on habitat preferences was obtained from Peterjohn (1989) and Peterjohn and Rice (1991).

ABUNDANCE AND DISTRIBUTION

Each species' abundance, based on average number of individuals recorded per route, was categorized as **very abundant** ($\geq$ 50.0), **abundant** (12.0 - 49.9), **common** (4.0 - 11.9), **fairly common** (2.0 - 3.9), **uncommon** (1.0 - 1.9), or **rare** (< 1.0). These quantitative definitions correspond well to the qualitative definitions of abundance used by Peterjohn et al. (1987). Each species' distribution, based on number of the 45 routes on which it was recorded, was categorized as **widely distributed** (40 - 45 routes), **fairly widely distributed** (30 - 39 routes), **locally distributed** (20 - 29 routes), or **very locally distributed** ($\leq$19 routes).

A distribution map, indicating relative abundance on each route, is given for each species. To facilitate between-species comparisons, we used a single set of symbol definitions for distribution maps of all species within a given abundance category. Similarly, a single Y-axis is used for species with similar abundance. Except for a few very abundant species, the maximum value on the Y-axis is either 0.5, 1, 3, 5, 10, 16, 30, or 40.

OHIO POPULATION TREND

Population trends were estimated by fitting an exponential curve (i.e., a curve having the same proportional change between every two consecutive years) to the annual mean of individuals reported per route. Thus, population trends are expressed as the annual percent change in the number of individuals recorded per year. The description "percent annual change ($\pm$ SE) = 5.6 ($\pm$ 2.3), P = 0.02" indicates that the population trend increased 5.6% per year, with a standard error of 2.3%, and a significance level of 0.02. Significance levels were based on a two-tailed t-test.

For those species that exhibited some annual means that were clear outliers from the general trend, the analysis was repeated with the outliers omitted. Likewise, for some species, the trend was exponential over only a portion of the 1966-1994 study interval and we restricted the curve-fitting to that period. Each species account includes a graph depicting the population trend line and the average number of individuals recorded per route in each year.

We ignored missing data in estimating trends and standard errors. Missing data have little affect on the analysis as long as (1) they comprise a small fraction of the potential data and (2) they are distributed widely across routes and years. The Ohio BBS data set meets both of these conditions. Ohio BBS routes have been well-covered except for a few years in the late 1980s (Figure 2). Ten routes were run in all 29 years, and most routes (53%) were run in $\geq$26 of the 29 years (Figure 3). Overall, missing data comprise only 194 (15%) of the 1305 potential route-years (i.e., 45 routes x 29 years) in the Ohio BBS data set.

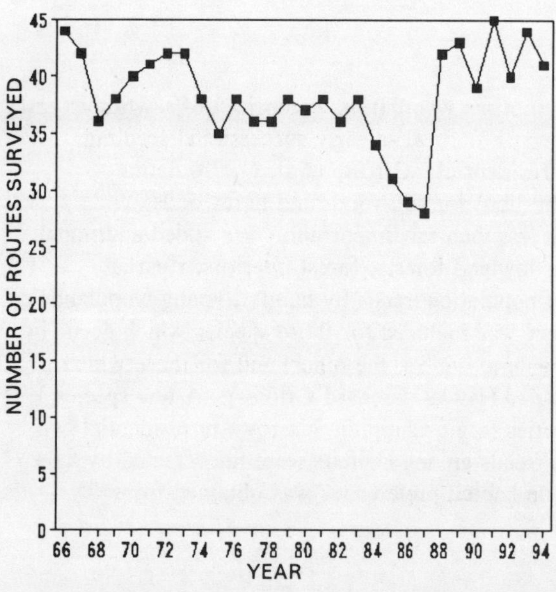

Fig. 2. Number of routes surveyed in each year.

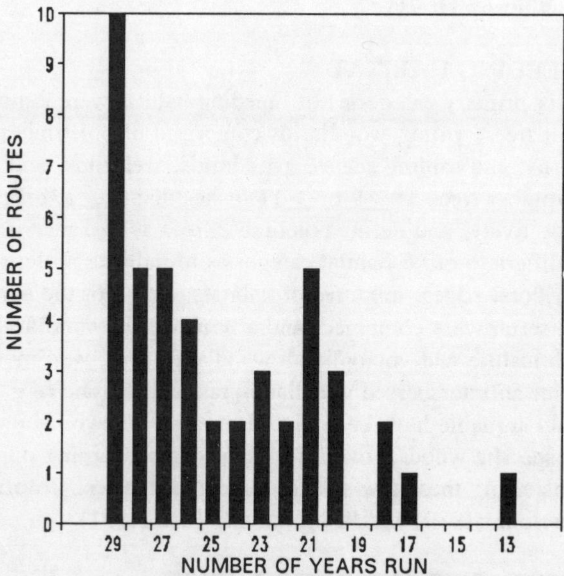

Fig. 3. Frequency distribution of number of years that each route was surveyed (maximum possible = 29).

Many different methods are currently being used to analyze BBS data and there is little agreement on which are better (e.g., Thomas, 1996). An advantage of our method is that the trend lines closely fit the annual means of individuals recorded per route, which in turn are assumed to reflect annual popolation size. Presumably this method has not been widely used in the past because an appropriate estimator of the variance was not available; thus, in conjunction with this study, Bart and Notz (in prep.) developed a variance estimator which appropriately acknowledges that the same routes are run each year and thus annual means are not entirely independent from one another. A more detailed description, including a comparison with the route-regression method used by researchers at the BBS Office (e.g., Geissler and Sauer, 1990; Link and Sauer, 1994), can be found in Appendix I.

CONTINENTAL AND GREAT LAKES REGIONAL TREND

The continental and regional population trends are given for each species. The continental trend was based on all BBS routes in the contiguous U.S. and Canada; the regional trend was based on all routes in the Great Lakes Region (Region 3) of the U.S. Fish & Wildlife Service which includes Minnesota, Iowa, Missouri, Wisconsin, Illinois, Indiana, Michigan, and Ohio. Continental and regional trends were supplied by the Breeding Bird Survey Office (Peterjohn, pers. comm.) and were based on the statistical method described by Geissler and Sauer (1990) and Sauer et al. (1994) (also see Appendix I). Note that the continental and regional analyses are based on more routes and thus have more statistical power than the Ohio analysis. Thus, even if the two populations are changing in a similar fashion, the former may produce a significant trend and the latter may not. One should conclude that the Ohio trend is different from the regional/continental trend only if both are significant and in opposite directions. Also, continental and regional trends should be interpreted with caution because complex patterns of increases and decreases in different periods or regions may cause the overall trend to be misleading. Thus, without a graphical presentation of the trend and an investigation into potentially opposing trends within the geographic region of interest (such as we present for the Ohio analysis), it is difficult to interpret regional/continental trends with confidence. Such an analysis is beyond the scope of this monograph. More detailed information on regional and continental trends can be obtained from the BBS office.

RESULTS AND DISCUSSION

ADEQUACY OF BBS FOR DETECTING POPULATION TRENDS OF OHIO'S BREEDING BIRDS

The North American Breeding Bird Survey is an excellent tool for monitoring long-term population trends in most of Ohio's breeding species. For some species, however, it is difficult to estimate a reliable population trend because they are consistently detected on few routes or in few years. Here we attempt to identify the types of species that are inadequately covered by BBS and thus those for which supplemental monitoring schemes might be desirable.

We use standard power calculations to define "adequate" data (Steel and Torrie, 1980). By our definition, a species is adequately covered in Ohio if the 29-year BBS data set provides a high probability ($\geq 80\%$) of detecting as significant a true population change of 5% or more; such species are those having a SE of the trend < 1.72. Note that the criteria used in a power equation are arbitrary. More stringent criteria, such as a higher probability threshold (e.g., 85%) or a lower population change threshold (e.g., 3%) would result in many more species being identified as inadequately covered. Our criteria are a compromise between the optimal desires of the manager and the realities of the Breeding Bird Survey. Our primary goal in this section is to identify the **types** of species that tend to be inadequately covered; this goal is most readily met with a definition such as ours that identifies an intermediate number of species as inadequately covered.

In general, BBS tends to be inadequate for those species that 1) occupy patchily distributed wetland habitat; 2) occupy forest interiors or patchily distributed forest types (e.g., riparian corridors, hemlock gorges, pine-dominated forests); 3) are fairly nonvocal during early morning hours; or 4) are inactive at times of day when BBS routes are run (Table 1). Such species, because of their behavior or habitat preference, tend to be sporadically recorded on BBS routes. Of the 9 species that do not readily fit into one of these categories (the grassland and scrub/nonspecific woodland species), 7 are rare and locally distributed species that are not often

Table 1. Ohio-breeding species for which BBS data is inadequate to estimate a reliable population trend (excludes the rare breeders listed in Appendix II). Inadequately covered species are those that had a SE of the population trend > 1.72 % (see text for further explanation).

WETLAND SPECIES	NOCTURNAL AND CREPUSCULAR SPECIES
Pied-billed Grebe (*Podilymbus podiceps*)	American Woodcock (*Scolopax minor*)
Least Bittern (*Ixobrychus exilis*)	Eastern Screech-Owl (*Otus asio*)
Great Blue Heron (*Ardea herodias*)[a]	Barred Owl (*Strix varia*)[a]
American Black Duck (*Anas rubripes*)	Whip-poor-will (*Caprimulgus vociferus*)
Mallard (*Anas platyrhynchos*)[a]	Common Nighthawk (*Chordeiles minor*)[a]
Blue-winged Teal (*Anas discors*)	
Hooded Merganser (*Lophodytes cucullatus*)	**OTHER RELATIVELY NON-VOCAL SPECIES**
Virginia Rail (*Rallus limicola*)	
Sora (*Porzana carolina*)	Red-shouldered Hawk (*Buteo lineatus*)[a]
Common Moorhen (*Gallinula chloropus*)	Broad-winged Hawk (*Buteo platypterus*)
American Coot (*Fulica americana*)	Ruffed Grouse (*Bonasa umbellus*)
Spotted Sandpiper (*Actitis macularia*)[a]	Wild Turkey (*Meleagris gallopavo*)
Common Snipe (*Gallinago gallinago*)	Brown Creeper (*Certhia americana*)
Herring Gull (*Larus argentatus*)	
Tree Swallow (*Tachycineta bicolor*)[a]	
Bank Swallow (*Riparia riparia*)[a]	**GRASSLAND SPECIES**
Cliff Swallow (*Hirundo pyrrhonota*)[b]	
Sedge Wren (*Cistothorus platensis*)	Vesper Sparrow (*Pooecetes gramineus*)[a]
Marsh Wren (*Cistothorus palustris*)	Henslow's Sparrow (*Ammodramus henslowii*)[a]
Prothonotary Warbler (*Protonotaria citrea*)	Upland Sandpiper (*Bartramia longicauda*)[a]
Swamp Sparrow (*Melospiza georgiana*)[a]	Western Meadowlark (*Sturnella neglecta*)[a]
FOREST PATCH SPECIALISTS[c]	
	SCRUB AND NONSPECIFIC WOODS SPECIES
Veery (*Catharus fuscescens*)	
Solitary Vireo (*Vireo solitarius*)	Least Flycatcher (*Empidonax minimus*)[a]
Northern Parula (*Parula americana*)	Chestnut-sided Warbler (*Dendroica pensylvanica*)
Black-throated Green Warbler (*Dendroica virens*)	Black-and-White Warbler (*Mniotilta varia*)[a]
Pine Warbler (*Dendroica pinus*)	Rose-breasted Grosbeak (*Pheucticus ludovicianus*)[a]
Worm-eating Warbler (*Helmitheros vermivorus*)	Blue Grosbeak (*Guiraca caerulea*)
Louisiana Waterthrush (*Seiurus motacilla*)[a]	
Hooded Warbler (*Wilsonia citrina*)[a]	
Purple Finch (*Carpodacus purpureus*)	

[a] Occurred on ≥ 14 routes and are therefore included in the Species Accounts.

[b] Cliff swallows also use upland habitats extensively.

[c] Species preferring forest interiors or patchily-distributed forest types such as hemlock gorges, pine-dominated stands, or forested riparian corridors.

recorded on BBS routes, and 2 (Vesper Sparrow and Rose-breasted Grosbeak) are well-distributed but have somewhat large SE's because their population trend differs greatly among routes. Not surprisingly, there is a close relationship between number of routes and years that a species has been recorded and the adequacy of BBS in monitoring that species. Nearly all species identified as inadequately covered (43 of 49) were observed in <30% of the route-years, and similarly, nearly all species that were observed in <30% of the route-years (43 of 48) were identified as inadequately covered.

Inadequately monitored species comprise 49 of the 137 breeding species recorded on BBS routes (excludes rare breeders, Appendix II) and include proportionately more non-passerines than passerines; 49% of non-passerines and only 29% of passerines breeding in Ohio are inadequately monitored by BBS.

To monitor Ohio population trends of the species listed in Table 1, it is necessary to supplement the BBS with other sampling schemes. For example, the Ohio Division of Wildlife (ODOW) has ongoing monitoring programs for the Barn Owl, Ruffed Grouse, and Wild Turkey, and in 1991, they initiated a Wetland Bird Survey (Andres, 1991) to monitor bitterns, herons, rails, wetland songbirds, and other wetland breeders. Also, the 9 nonrandom BBS routes recently added in Ohio were strategically placed to increase the sampling of forest interior species.

An additional, it may be advisable to monitor population trends of inadequately covered species (and others) on a regional basis (i.e., by including data from ecologically similar patches in adjacent states). A regional approach would provide the manager with much more statistical power to identify species that are declining at less than 5% annually. For example, it would be beneficial to have a high probability of detecting those species that are declining at 2.4% annually, a rate of decline that will halve the population over the course of the 29-year BBS data set. (The annual rate of change that we use, 5%, is so severe that the population would be decreased by one half in only 13.5 years.) However, under the 2.4% criterion, Ohio routes alone provide adequate data for only 41 of the 137 breeding species. A regional approach, on the other hand, is likely to provide enough data for most of the 137 species and may also be sufficient for many species that rarely breed in Ohio (Appendix II) but that may be of special interest in Ohio or the region.

POPULATION TRENDS

The population trends of Ohio's breeding birds are discussed in detail in the following Species Accounts. An analysis of trends by migratory status and breeding habitat is presented below. The analysis is based on 98 of the 105 species included in the Species Accounts; it excludes the 7 species that exhibited strongly non-exponential trends due to sharp declines after the severe winters of 1976-1977 and 1977-1978. Note that the various methods of analysis currently in use often give different results (e.g., Thomas, 1996 and James et al. 1996). A comparison of results based on the method used here, a method incorporating observer effects, and the method used by researchers in the BBS Office is included in Appendix I.

Overall, 21% of Ohio's breeding birds are declining significantly and 45% are increasing significantly. The significantly declining species, in taxonomic order, were the Ring-necked Pheasant, Northern Bobwhite, Upland Sandpiper, Black-billed Cuckoo, Yellow-billed Cuckoo, Red-headed Woodpecker, Northern Flicker, Least Flycatcher, Great Crested Flycatcher, Purple Martin, Yellow-breasted Chat, Summer Tanager, Dickcissel, Field Sparrow, Grasshopper Sparrow, Henslow's Sparrow, Bobolink, Red-winged Blackbird, Eastern Meadowlark, Western Meadowlark, and House Sparrow.

Neotropical migrants did not appear to be declining proportionately more than temperate migrants or residents; 11 neotropical migrants (23%) and 10 temperate migrants or residents (20%) exhibited declines. Numbers of significantly increasing species were also similar in the two migratory categories (20 neotropical (43%) vs. 24 temperate migrants and residents (47%)).

Grassland breeding species exhibited proportionately more declines than did species breeding in other habitats (Tables 2 and 3). Ten of the 14 (71%) grassland species declined significantly and the ratio of significantly declining to increasing species was 10:1. Mature woodlands had a much lower ratio of significantly declining to increasing species (1:6) than other habitat categories. Intermediate percentages of species breeding in nonspecific woodland, residential, and scrub/young woodland habitats exhibited significant declines (19, 14, and 11%, respectively). Unfortunately, it is difficult to compare wetland breeding species to those of other habitats, because 1) wetland species are not well-represented in BBS data, and 2) the well-represented species (e.g., Mallard, Canada Goose) may be a biased sample of wetland species because they tend to be less impacted by human activities, and thus less likely to be decreasing, than the other more secretive species (e.g., bitterns, rails). It is also difficult to draw conclusions about species breeding in residential areas because only 7 species are included in this category.

Table 2. Percent of species exhibiting significantly decreasing and increasing trends (1966-1994).

Breeding Habitat[a] (Total no. of species)	Percent declining	Percent increasing	Ratio of declining to increasing species
Grasslands (14)	71	7	10:1
Nonspecific Woods[b] (27)	19	59	1:3
Scrub/Young Woods (18)	11	39	1:3
Mature Woods (20)	10	60	1:6

[a]Of the 12 wetland breeding species, 8% were declining and 33% increasing; of the 7 species breeding in residential areas, 14% were declining and 57% increasing.
[b]Includes species occurring in both mature and scrub/young woods (i.e., not easily assigned to one category or the other), and species occurring in a forest-grassland mosaic (e.g., Red-tailed Hawk and Orchard Oriole).

Some patterns of population change can be explained by weather. The severe winters of 1976-1977 and 1977-1978 caused marked declines in several residents and local migrants: Northern Mockingbird, Carolina Wren, Mourning Dove, Common Grackle, Tufted Titmouse, Eastern Bluebird, Brown-headed Cowbird, and Northern Bobwhite. In the 15 years since the severe winters, all species, except Common Grackle and Northern Bobwhite, have returned to or above pre-1977 levels. Interestingly, winter effects were not obvious in some other residents such as the Black-capped or Carolina Chickadee, Hairy or Red-bellied Woodpecker, White-breasted Nuthatch, and Blue Jay.

In summary, our results from Ohio BBS clearly support the recent realization that grassland breeding species have incurred more drastic declines than other groups (e.g., Knopf, 1994), and that, at least in some geographic areas, neotropical migrants are not declining proportionately more than temperate migrants or residents (see also Hagan et al., 1992; and Bohning-Gaese et al., 1993) and species breeding in mature woodland are not declining more than those in other forest habitats (see also James et al., 1992; and Witham and Hunter, 1992). Because ecologically sound management within Ohio should be based, in part, on regional population trends and regional trends in habitat loss and availability, it would be interesting to perform a detailed analysis such as ours on data that includes ecologically similar regions in adjacent states and perhaps throughout each species' range. By facilitating comparisons among areas with different management regimes (e.g., areas where mature forest has increased vs. decreased), such an approach may also help to illimunate potential causes of declines for some species.

Table 3. Species exhibiting significantly decreasing or increasing trends (1966-1994). Asterisks indicate statistical significance of the population trend ($*$ $p < 0.05$; $**$ $p < 0.01$; $***$ $p < 0.001$). Neotropical migrants shown in bold.

Breeding Habitat[a]	Significantly Declining Species[b] (% annual change)	Significantly Increasing Species[b] (% annual change)	
Grasslands (14)	Western Meadowlark (-10.8)*** **Upland Sandpiper (-10.5)**** Northern Bobwhite (-8.1)*** Henslow's Sparrow (-6.2)* **Dickcissel (-5.9)**** Ring-necked Pheasant (-5.3)*** **Grasshopper Sparrow (-4.8)***** Eastern Meadowlark (-3.5)*** **Bobolink (-2.0)*** Field Sparrow (-1.2)*	Killdeer (4.4)***	
Nonspecific Woods[c] (27)	**Purple Martin (-5.2)***** Red-headed Woodpecker (-4.7)*** **Yellow-billed Cuckoo (-3.1)***** **Black-billed Cuckoo (-3.0)*** Northern Flicker (-2.0)***	Cedar Waxwing (6.6)*** **Ruby-throated Hummingbird (6.3)***** Black-capped Chickadee (6.2)*** White-breasted Nuthatch (5.0)*** Red-tailed Hawk (4.4)*** Chipping Sparrow (4.1)*** American Kestrel (3.9)*** Carolina Chickadee (3.9)***	Orchard Oriole (3.8)* American Robin (2.9)*** Blue Jay (2.9)*** Red-bellied Woodpecker (2.9)*** **Red-eyed Vireo (2.8)***** **Wood Thrush (2.0)***** Downy Woodpecker (2.0)* American Crow (1.6)***
Scrub or Young Woods (18)	**Least Flycatcher (-5.6)*** **Yellow-breasted Chat (-3.0)*****	**White-eyed Vireo (4.5)***** House Wren (4.2)*** **Gray Catbird (3.4)***** **Common Yellowthroat (2.6)*****	Northern Cardinal (2.5)*** **Indigo Bunting (2.4)***** American Goldfinch (1.3)*
Mature Woods (20)	**Summer Tanager (-4.5)*** Great Crested Flycatcher (-1.5)*	Cooper's Hawk (6.5)*** **Hooded Warbler (5.1)*** **Ovenbird (4.6)**** **Yellow-throated Vireo (4.1)***** Great Horned Owl (3.9)* Turkey Vulture (3.1)**	**Scarlet Tanager (2.8)***** **Acadian Flycatcher (2.8)*** **Kentucky Warbler (2.8)*** **Yellow-throated Warbler (2.5)**** Pileated Woodpecker (2.4)*** **Blue-Gray Gnatcatcher (1.9)***
Wetlands (12)	Red-winged Blackbird (-4.5)***	Canada Goose (17.8)*** Tree Swallow (6.5)*** Great Blue Heron (5.6)* **Northern Rough-winged Swallow (4.0)*****	
Residential (7)	House Sparrow (-2.5)***	House Finch[d] (57.3)*** **Common Nighthawk (6.1)*** **Barn Swallow (2.5)***** **Chimney Swift (1.3)***	

[a]Total number of species in each habitat given in parentheses.

[b]The order of species from most to least severely declining is not based on statistical comparisons of their population trends, and thus, any conclusions based on the precise ordering presented here should be made with caution.

[c]Includes species occurring in both mature and scrub/young woods (i.e., not easily assigned to one category or the other), and species occurring in a forest-grassland mosaic (e.g., Red-tailed Hawk and Orchard Oriole).

[d]Percent annual change calculated for 1980-1994; first occurred on a BBS route in 1980.

SPECIES ACCOUNTS

GREAT BLUE HERON

Ardea herodias

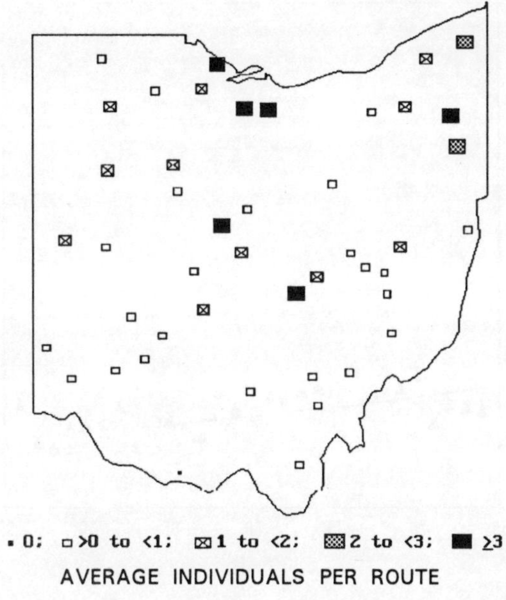

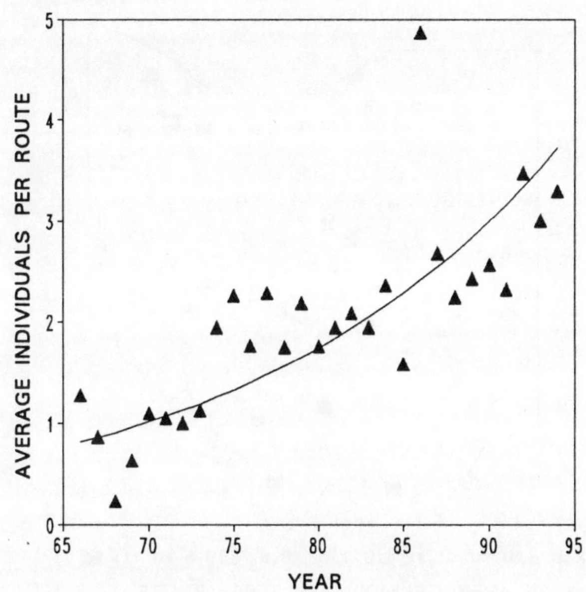

MIGRATORY STATUS: Temperate migrant.

BREEDING HABITAT: Tall mature trees for nesting and wetlands for foraging.

ABUNDANCE AND DISTRIBUTION: Uncommon (1.9 birds per route) but widely distributed (44 routes). Equally abundant in Western and Eastern Ohio (2.2 vs. 1.6 birds per route), but significantly more common in Northern than Southern Ohio (3.2 vs. 0.8, P = 0.02).   The largest heronries are along western Lake Erie where more than 1000 pairs nest on West Sister Island; most inland heronries have fewer than 75 pairs (Peterjohn and Rice, 1991). Peterjohn (1989) estimates Ohio's population at over 5000 breeding pairs.

OHIO POPULATION TREND: **Percent Annual Change (± SE) = 5.6 (± 2.3), P = 0.02**
   Great Blue Herons increased significantly at 5.6% annually. The trend remains significant when the outlier in 1986 is removed.
   The population increase was somewhat, but not significantly, greater in Southern than Northern Ohio (9.8 vs.4.8% annually, P = 0.09), suggesting that inland heronries are growing more rapidly than those along Lake Erie. Similarly, the increase was somewhat, but not significantly, greater in Eastern than Western Ohio (9.5 vs. 3.7%, P = 0.10).
   Great Blue Herons have probably been increasing throughout the twentieth century. During recent years, their distribution has expanded in the eastern glaciated counties and along the Scioto River, but has contracted in western counties since the 1950s (Peterjohn and Rice, 1991).

CONTINENTAL AND GREAT LAKES REGIONAL TREND: The continental and regional populations also increased significantly at 3.0 and 4.3% annually.

GREEN-BACKED HERON

Butorides striatus

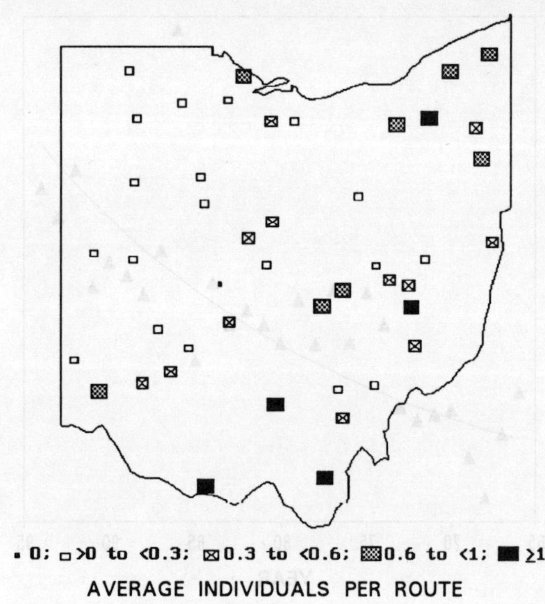

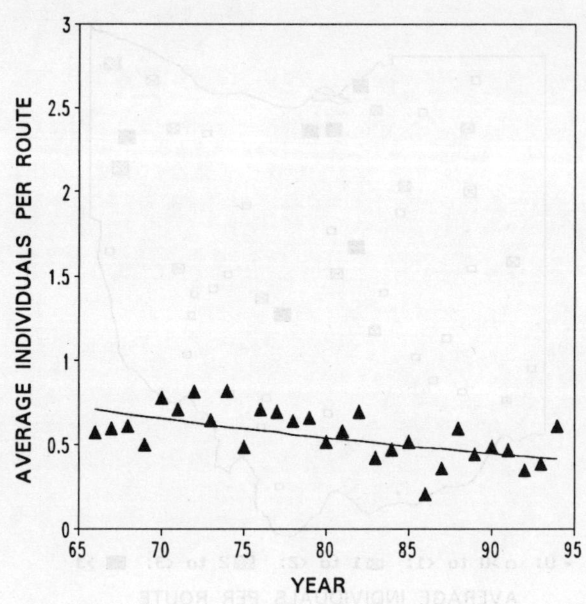

MIGRATORY STATUS: Central neotropical migrant.

BREEDING HABITAT: Wetlands with brushy thickets around margin, including streams, rivers, lakes, and marshes.

ABUNDANCE AND DISTRIBUTION: Rare (0.55 birds per route) but widely distributed (44 routes). More common in Eastern than Western Ohio (0.74 vs. 0.39, P = 0.007).

OHIO POPULATION TREND: **Percent Annual Change (± SE) = -1.9 (± 1.1), P = 0.09**
   Green-backed Herons exhibited a marginally significant decline of 1.9% annually. Green-backed Herons declined significantly in Western Ohio at 4.6% annually (P = 0.002), where they are less common, but did not exhibit a significant trend in Eastern Ohio (-0.4%, P = 0.75). The trends were significantly different (P = 0.03).

CONTINENTAL AND GREAT LAKES REGIONAL TREND: Neither the Great Lakes nor continental population exhibited a significant overall trend.

WOOD DUCK

Aix sponsa

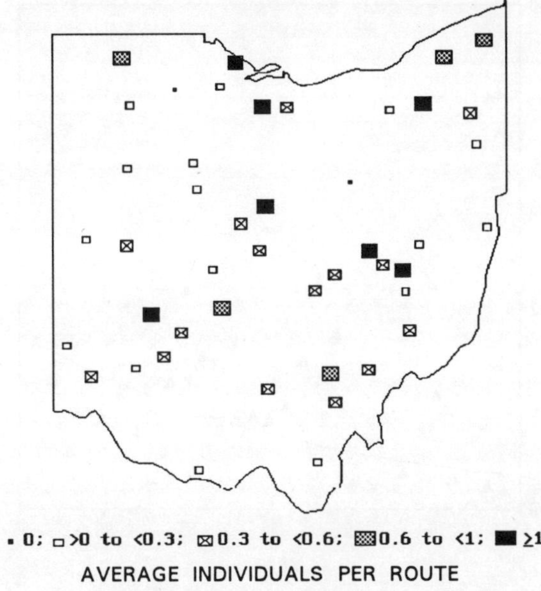
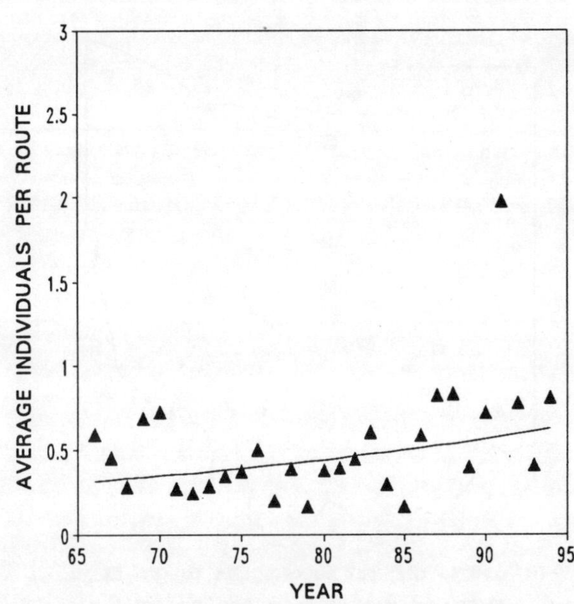

MIGRATORY STATUS: Temperate migrant.

BREEDING HABITAT: Wetlands, especially wooded swamps, marshes, and streams.

ABUNDANCE AND DISTRIBUTION: Rarely recorded on BBS routes (0.53 birds per route) but widely distributed (43 routes). Equally rare in Eastern and Western Ohio (0.54 vs. 0.53 birds per route, P = 0.90).

OHIO POPULATION TREND: Percent Annual Change (± SE) = 2.5 (± 1.5), P = 0.10

Wood Ducks increased somewhat, but not significantly, at 2.5% annually. The high count in 1991 is due to two large flocks observed on eastern routes (routes 48 and 51). Removing the flocks makes the overall trend even less significant (2.1%, P = 0.17). The trend is nearly significant in Eastern Ohio (4.2%, P = 0.06) even when the flocks are removed, and the population in Western Ohio is fairly stable (1.4%, P = 0.50).

After declining in the early 1900s due to overharvesting and loss of most of Ohio's wooded wetlands, Wood Ducks began to increase in the 1940s when the practice of providing nest boxes became popular, and they are now one of Ohio's most common and widespread waterfowl (Peterjohn, 1989).

CONTINENTAL AND GREAT LAKES REGIONAL TREND: Wood Ducks have increased significantly in the region and continent-wide at 5.2 and 6.1% annually.

MALLARD

Anas platyrhynchos

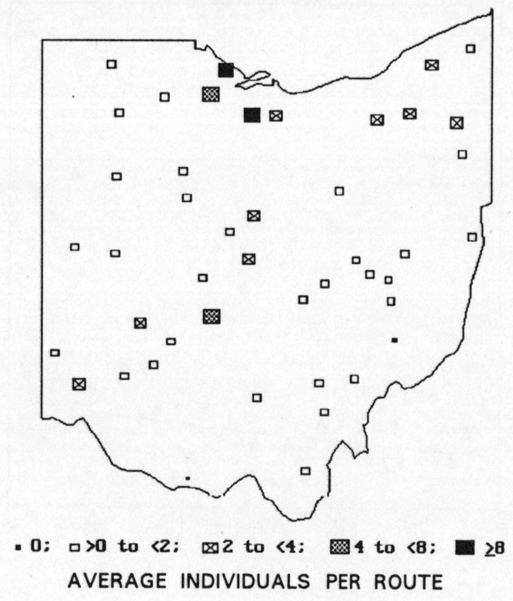

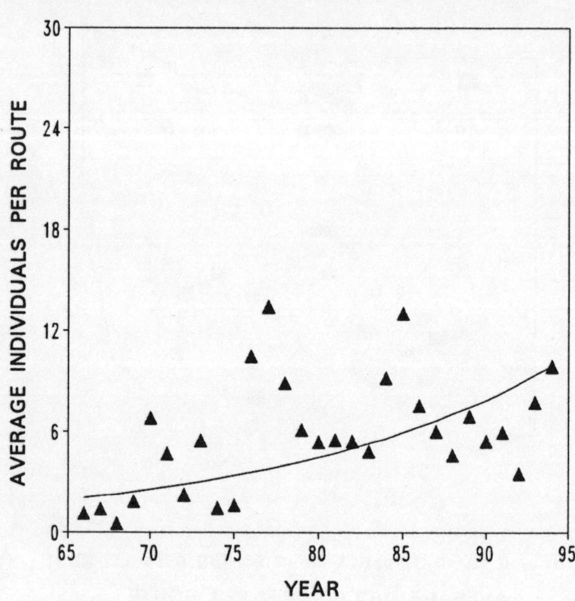

MIGRATORY STATUS: Temperate migrant, although thousands winter in the Lake Erie marshes.

BREEDING HABITAT: Wetlands, including small and highly modified wetlands such as farm ponds, ditches, and reservoirs.

ABUNDANCE AND DISTRIBUTION: Common (5.2 birds per route) and widely distributed (43 routes). Particularly common on two routes in the Lake Erie Marshes which averaged 46 and 136 birds per route; all other routes averaged <6 birds per route.

An increase in Ohio's Mallard population and an expansion into southern Ohio became apparent during the 1940s and 1950s (Peterjohn, 1989) and has continued during the last several decades.

OHIO POPULATION TREND: Percent Annual Change ($\pm$ SE) = 5.8 ($\pm$ 3.4), P = 0.09

Mallards increased substantially at 5.8% annually, but the trend was not significant due to large annual variation. The increase, despite the loss of natural wetland habitat, probably results from the Mallard's ability to adapt to a diversity of habitats, including those modified by humans.

Trends in Western vs. Eastern (5.6 vs. 8.5%, P = 0.51) and Northern vs. Southern Ohio (5.8 vs. 9.1%, P = 0.44) did not differ signficantly.

CONTINENTAL AND GREAT LAKES REGIONAL TREND: Mallards have increased significantly in the region and continent-wide at 5.1 and 2.0% annually.

CANADA GOOSE

Branta canadensis

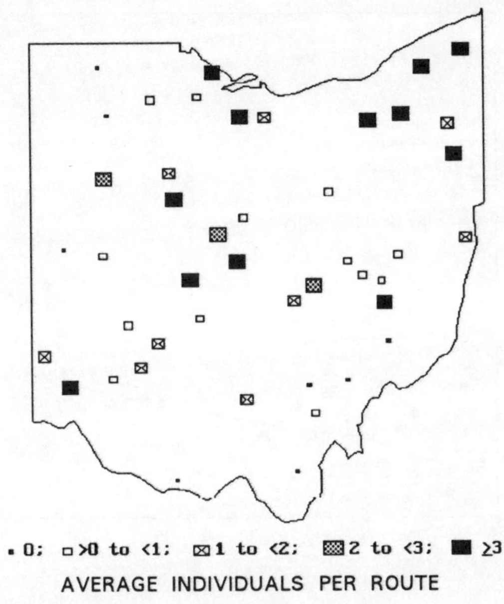

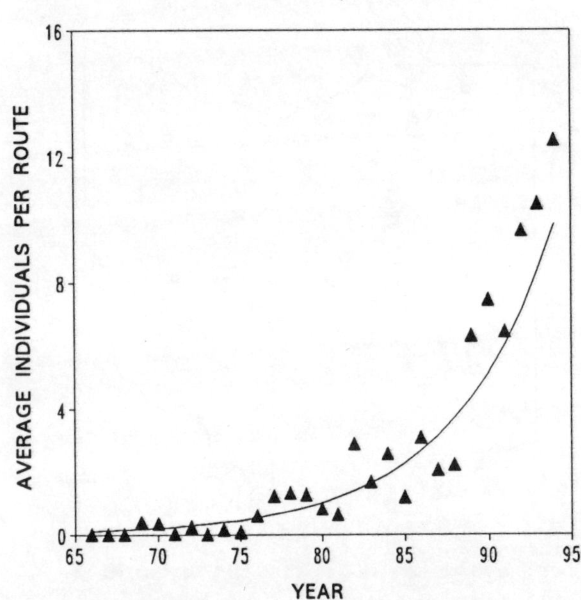

MIGRATORY STATUS: Permanent resident.

BREEDING HABITAT: Wetlands.

ABUNDANCE AND DISTRIBUTION: Fairly common (2.6 birds per route) and fairly widely distributed (37 routes). More common in Northern than Southern Ohio (4.0 vs. 1.3 birds per route, P = 0.02).

OHIO POPULATION TREND: Percent Annual Change ($\pm$ SE) = 17.8 ($\pm$ 1.2), P < 0.001
  Ohio's resident Canada Goose population has increased at a striking 17.8% annually since 1966 and has increased particularly dramatically since 1980. The increase has been similar in Eastern vs. Western (16.3 vs. 15.8, P = 0.78) and Northern vs. Southern Ohio (16.8 vs. 16.7%, P = 0.96).
  Although historically Canada geese rarely bred in Ohio, state agencies began to introduce breeding, resident pairs to Ohio in the early 1950's in an attempt to restore the continental population which had been reduced by overhunting in the preceding decades (Peterjohn, 1989). Initially the reintroduced populations remained on state-managed lands but began rapidly spreading throughout the state in the late 1970s. Their spread was facilitated by their ability to adapt to small and disturbed wetland breeding habitat, including those in urban areas.

CONTINENTAL AND GREAT LAKES REGIONAL TREND: Similarly, the Canada Goose has increased significantly in the region and continent-wide at 23.2 and 11.7% annually.

TURKEY VULTURE

Cathartes aura

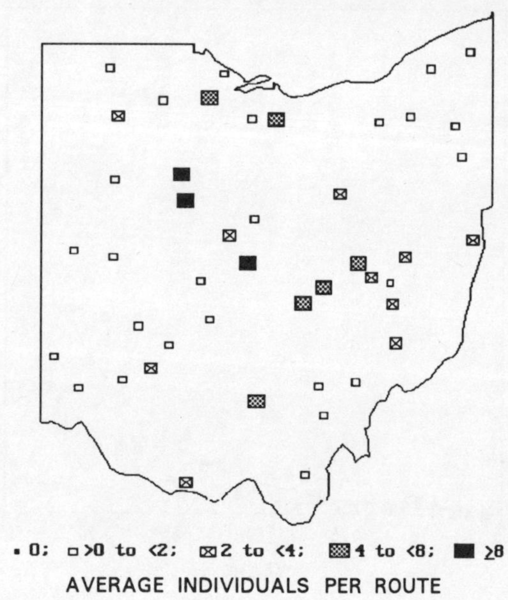

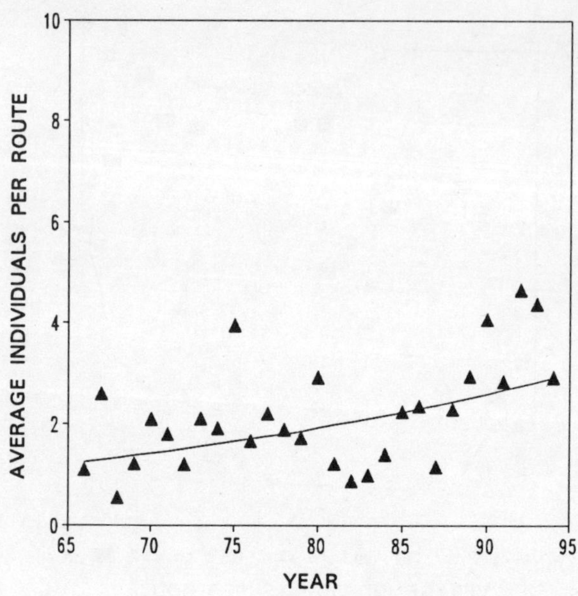

MIGRATORY STATUS: Temperate migrant, a few winter in southern Ohio.

BREEDING HABITAT: Extensive woodlands or on rocky cliffs and ledges.

ABUNDANCE AND DISTRIBUTION: Fairly common (2.2 birds per route) and widely distributed (all routes). Turkey Vultures are probably underrepresented by BBS because routes are run in the early morning before thermals have developed and thus at a time when vultures are relatively inactive.

Equally abundant in Western and Eastern Ohio (2.3 vs. 2.1 birds per route, respectively).

OHIO POPULATION TREND: **Percent Annual Change ($\pm$ SE) = 3.1 ($\pm$ 1.1), P = 0.006**

Turkey Vultures have increased significantly at 2.8% annually since 1966 despite large annual variation in counts. The large annual variation probably results from the tendency of vultures to occur in flocks which are sparsely distributed throughout Ohio and which vary in distribution from year to year. Trends in Western and Eastern Ohio were similar (3.3 vs. 3.1% annual change, P = 0.92).

The increase in Ohio may result, in part, from the continental population's range expansion into the Great Lakes region (Robbins et al., 1986).

CONTINENTAL AND GREAT LAKES REGIONAL TREND: The continental population also increased significantly at 0.9% annually but not as greatly as in Ohio and the region (3.4%).

COOPER'S HAWK

Accipiter cooperii

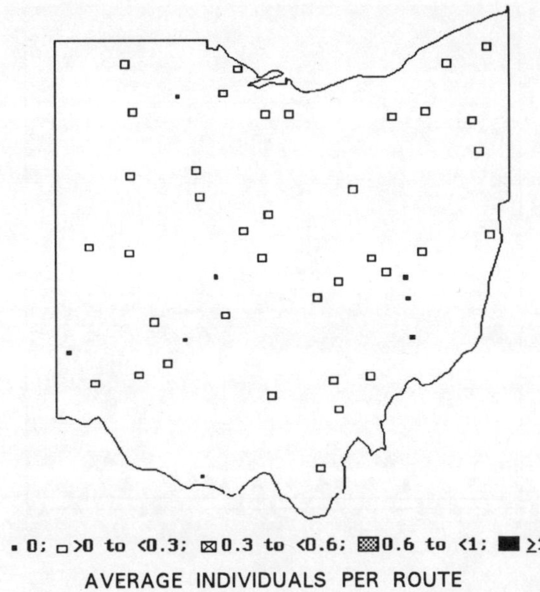

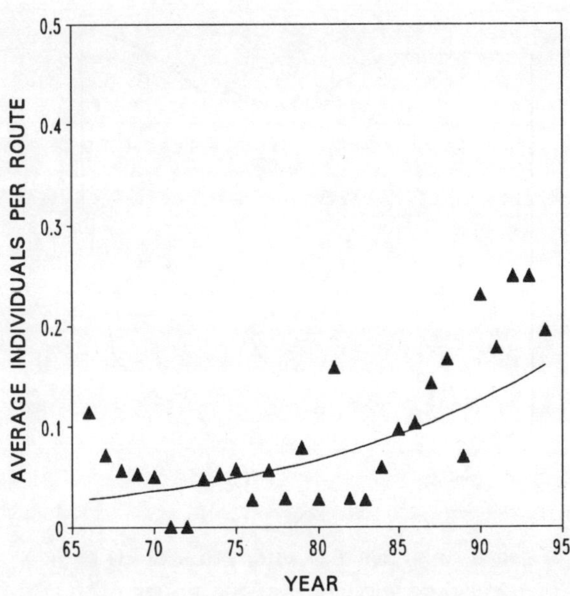

MIGRATORY STATUS: Permanent resident or temperate migrant. Many winter in Ohio; it is not known whether individuals that breed in Ohio also winter there.

BREEDING HABITAT: Mature woodlands, especially those with dense cover.

ABUNDANCE AND DISTRIBUTION: Rarely recorded overall (0.10 birds per route) but fairly widely distributed (37 routes). Equally abundant in Western and Eastern Ohio (0.09 vs. 0.10 birds per route, P = 0.62). Cooper's Hawks are probably underrepresented in BBS data because they initiate nests prior to BBS data collection in early June and are not very vocal thereafter.

OHIO POPULATION TREND: Percent Annual Change (± SE) = 6.5 (± 1.3), P < 0.001

Overall the population has increased since 1966 at a significant rate of 6.5% annually. The increase has been particularly noticeable since 1980. The increase was somewhat, but not significantly, higher in Western than Eastern Ohio (8.3 vs. 4.6% annually, P = 0.16).

Ohio's population began declining in the 1940s, probably due to contamination from DDT and other pesticides, and its recent recovery, beginning in the early 1970s, is believed to result from the ban of those pesticides (Peterjohn, 1989).

CONTINENTAL AND GREAT LAKES REGIONAL TREND: The continental and regional populations have also increased at 5.5 and 5.3%, respectively, although only the continental trend is statistically significant. The continent-wide decline and subsequent recovery presumably also result from the use and ban of DDT and related pesticides (Robbins et al., 1986).

SHARP-SHINNED HAWK

Accipiter striatus

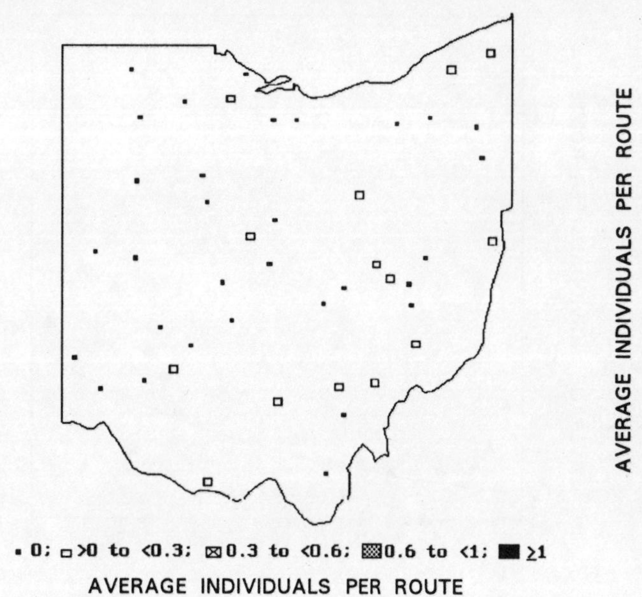

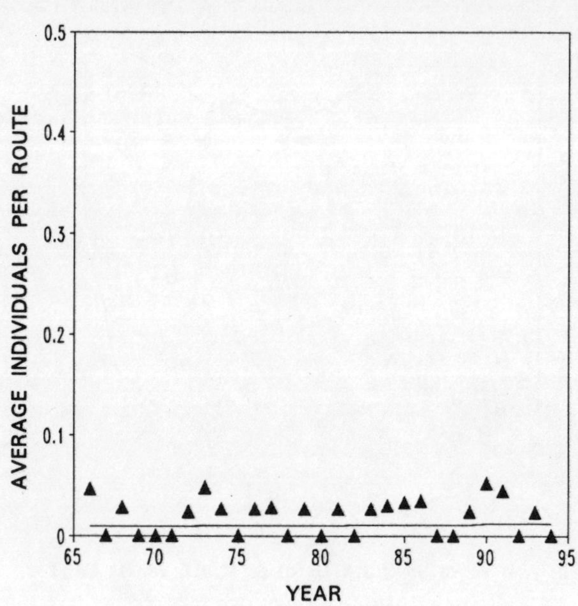

MIGRATORY STATUS: Temperate and central neotropical migrant.

BREEDING HABITAT: Mature and young woodlands, often with conifers.

ABUNDANCE AND DISTRIBUTION: Rare (0.02 birds per route) and very locally distributed (14 routes). More common in Eastern than Western Ohio (0.03 vs. 0.01 birds per route, P = 0.05).

OHIO POPULATION TREND: Percent Annual Change ($\pm$ SE) = 1.0 ($\pm$ 1.1), P = 0.40

Sharp-shinned Hawks have not exhibited a significant overall trend (1.0% annual change). Any trend would be difficult to detect due to the small number of individuals observed per route-year. Distribution inadequate to compare trends in Eastern vs. Western Ohio.

Thought to have declined in Ohio and elsewhere due to DDT and other pesticides and to have recovered quickly after those pesticides were banned (Peterjohn, 1989).

CONTINENTAL AND GREAT LAKES REGIONAL TREND: Neither the regional nor continental populations exhibited a significant overall trend (3.9 and 1.4% annual change).

RED-TAILED HAWK

Buteo jamaicensis

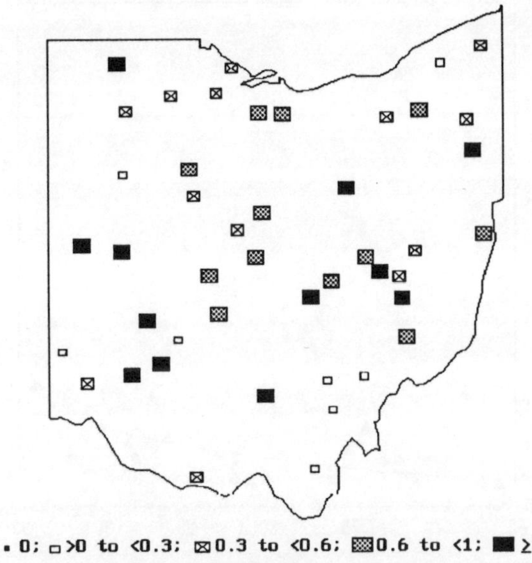

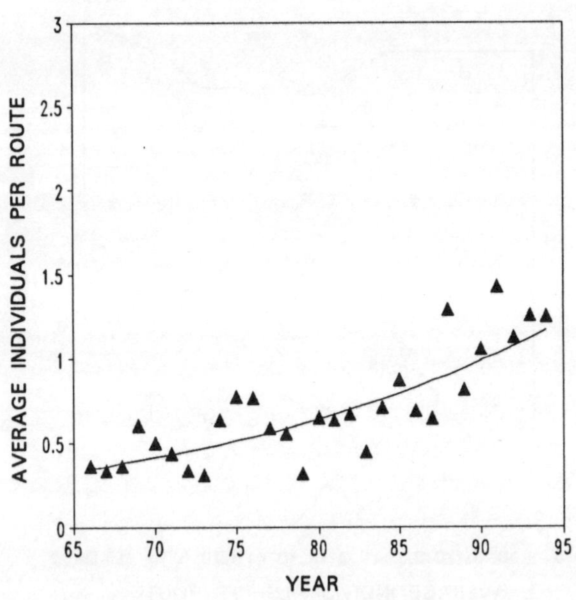

MIGRATORY STATUS: Permanent resident or temperate migrant. Most individual breeding pairs are thought to winter in Ohio (Peterjohn, 1989).

BREEDING HABITAT: Mature woodlands with adjacent grasslands.

ABUNDANCE AND DISTRIBUTION: Rarely recorded (0.74 birds per route) but widely distributed (all routes). Like other raptors, Red-tailed Hawks are probably underrepresented in BBS data, relative to other species. However, Red-taileds are abundant enough that BBS data provide a good estimate of trends through time.

Equally abundant in Western and Eastern Ohio (0.73 vs. 0.74 birds per route, P = 0.90).

OHIO POPULATION TREND: Percent Annual Change ($\pm$ SE) = 4.4 ($\pm$ 0.7), P < 0.001

The Ohio population has increased significantly since 1966 at 4.4% annually. The observed increase is part of the gradual increase that began in the 1940s (Peterjohn, 1989) and is probably due to an increase in mature forest habitat. The increasing trend was similar in Western and Eastern Ohio (4.0 vs. 5.1% annually, P = 0.45).

CONTINENTAL AND GREAT LAKES REGIONAL TREND: The continental and regional populations also increased significantly at 3.2% annually.

RED-SHOULDERED HAWK

Buteo lineatus

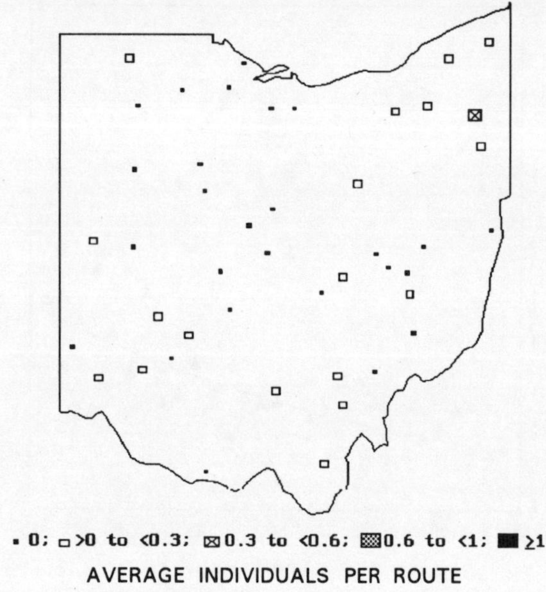

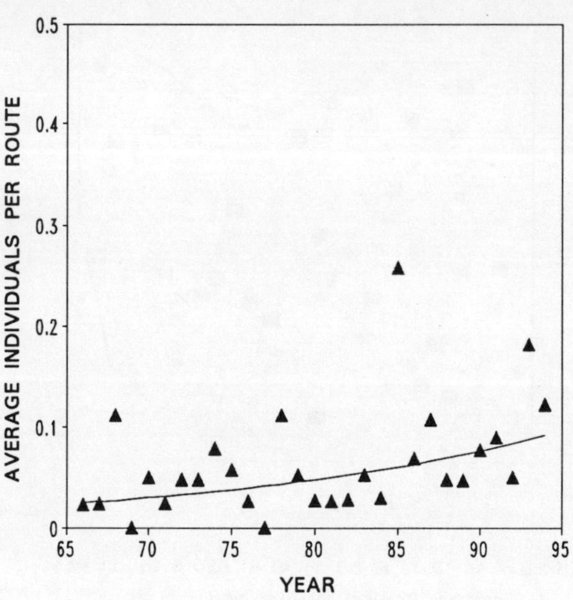

MIGRATORY STATUS: Temperate migrant.

BREEDING HABITAT: Mature woodlands, especially those in riparian areas.

ABUNDANCE AND DISTRIBUTION: Rarely recorded overall (0.06 birds per route) and very locally distributed (19 routes). Like Cooper's Hawks, Red-Shouldered Hawks initiate nests prior to BBS data collection, are not very vocal after initiation, and thus are detected at a low rate relative to other species (e.g., songbirds). The low detection rate contributes to the large year-to-year variation in number recorded per route.

   Less common in Western than Eastern Ohio (0.02 vs. 0.11 birds per route, P = 0.01). The near disappearance of Red-shouldereds in northwestern and northcentral counties occurred between 1940 and 1970, was due to destruction of riparian forests, and was accompanied by an increase in many southwestern and eastern counties (Peterjohn, 1989).

OHIO POPULATION TREND: Percent Annual Change (± SE) = 4.8 (± 2.8), P = 0.09

   The 4.8% annual increase in Ohio's population was not statistically significant, due in part to the substantial year-to-year variation in number of Red-shouldereds recorded per year. Red-Shouldered Hawks increased significantly in Eastern Ohio at 7.3% annually (P = 0.01), but occurred on too few routes in Western Ohio to estimate the trend with reasonable precision.

CONTINENTAL AND GREAT LAKES REGIONAL TREND: Like Ohio's population, the regional population increased (2.4% annual change), but not significantly. The continental population increased significantly at 1.6% annually.

AMERICAN KESTREL

Falco sparverius

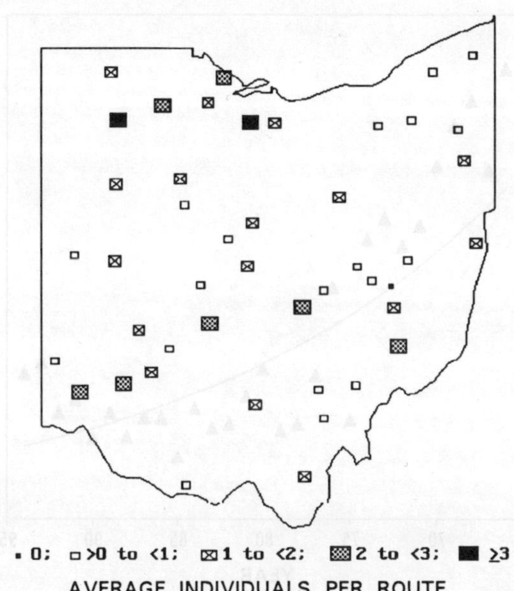

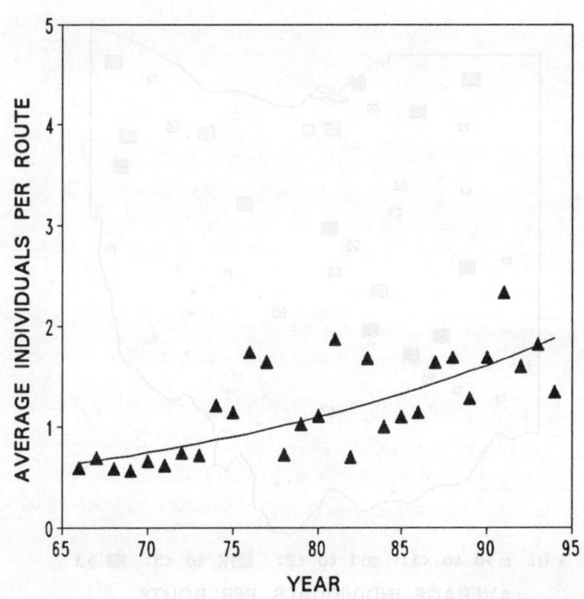

MIGRATORY STATUS: Permanent resident, although there is much winter movement.

BREEDING HABITAT: Mature woodlands with adjacent grasslands.

ABUNDANCE AND DISTRIBUTION: Uncommonly recorded (1.2 birds per route) and widely distributed (44 routes). Like other raptors, kestrels are probably underrepresented in BBS data, relative to other species. However, like Red-tailed Hawks, American Kestrels are abundant enough that BBS data provide a good estimate of trends through time.

American Kestrels are significantly more common in Western than Eastern Ohio (1.5 vs. 0.8 birds per route, P = 0.006).

OHIO POPULATION TREND: Percent Annual Change (± SE) = 3.9 (± 1.0), P < 0.001

American Kestrels have increased significantly since 1966 at 3.9% annually. The increasing trend was similar in Western and Eastern Ohio (4.3 vs. 3.4%, P = 0.64). Kestrels probably began to increase as settlers cleared virgin forests and have continued to adapt well to land-use changes in Ohio (Peterjohn, 1989).

CONTINENTAL AND GREAT LAKES REGIONAL TREND: American Kestrels in the Great Lakes Region have also increased significantly at 2.4% annually, but the Eastern Region and continental populations have not exhibited significant trends (0.5 and 0.3%).

RING-NECKED PHEASANT

Phasianus colchicus

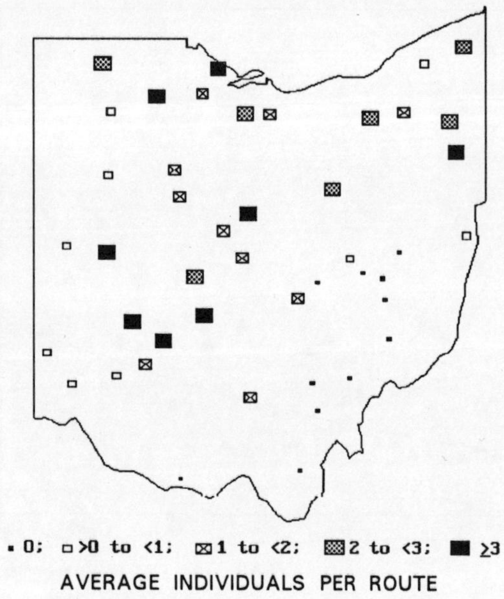

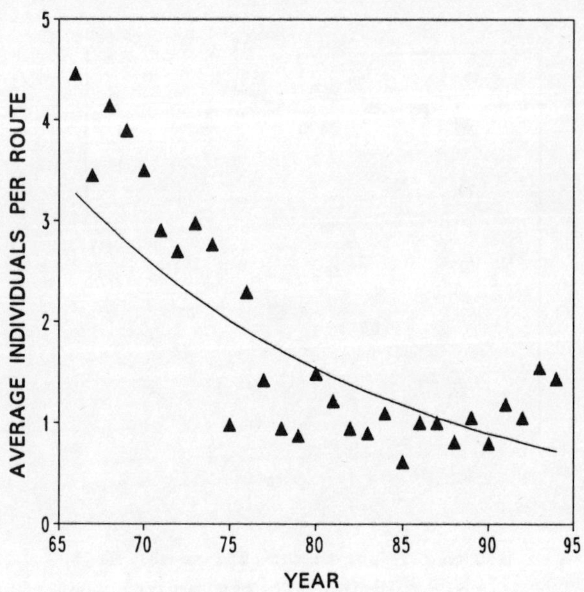

MIGRATORY STATUS: Permanent resident.

BREEDING HABITAT: Grasslands.

ABUNDANCE AND DISTRIBUTION: Fairly common (2.0 birds per route) and fairly widely distributed (34 routes). More common in the grasslands of Western Ohio than in Eastern Ohio (2.9 vs. 0.9, P = 0.01). Not present in the Unglaciated Plateau of southeastern Ohio.

OHIO POPULATION TREND: Percent Annual Change (± SE) = -5.3 (± 0.7), P < 0.001
  Ohio's Ring-necked Pheasant population has declined significantly since 1966 at 5.3% annually. The decline had occurred by the late 1970s and the population has remained fairly stable since then. Declines were similar in Western and Eastern Ohio (-5.3 vs. -6.4% annual change, P = 0.62).
  After large numbers were introduced throughout Ohio during 1913-1935, Ring-necked Pheasants quickly disappeared from eastern counties with unsuitable habitat, and have been declining statewide since the mid-1940's despite annual releases (Peterjohn, 1989). Efforts to stock pheasants are exacerbated by the continued loss of suitable grassland habitat in western counties.

CONTINENTAL AND GREAT LAKES REGIONAL TREND: Similarly, Ring-necked Pheasants have decreased significantly in the region and continent-wide at 2.1 and 1.2% annually.

NORTHERN BOBWHITE

Colinus virginianus

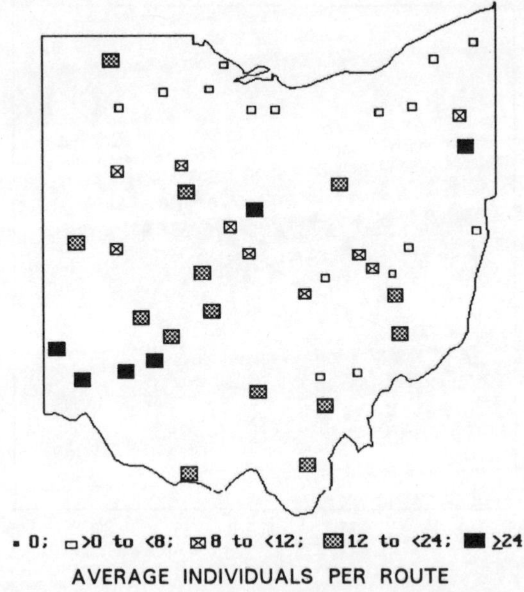

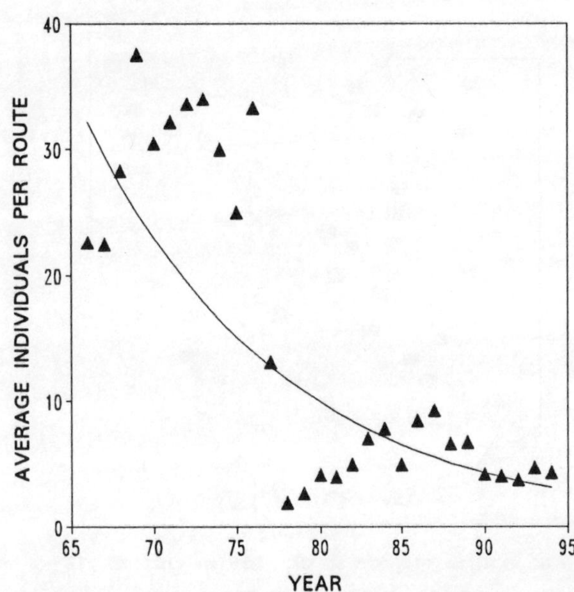

MIGRATORY STATUS: Permanent resident.

BREEDING HABITAT: Grasslands interspersed with brush.

ABUNDANCE AND DISTRIBUTION: Abundant (15.0 birds per route) and widely distributed (all routes). More common in the grasslands of Western Ohio than in Eastern Ohio (18.6 vs. 10.9 birds per route, P = 0.04). Ohio is on the northern edge of the bobwhite's range, and thus, they are more common in Southern than Northern Ohio (19.6 vs. 9.7 birds per route, P = 0.007).

OHIO POPULATION TREND: Percent Annual Change (± SE) = -8.1 (± 1.4), P < 0.001

Since 1966, Northern Bobwhite have declined at an overall rate of 8.1% annually. The population was severely reduced during the extreme winters of 1976-1977 and 1977-1978, rebounded during 1978-1987 (2.1%, P < 0.001) and has declined significantly since 1987 at 10.1% annually (P < 0.001). Although the decline is due to severe winter weather, the Northern Bobwhite's inability to rebound to earlier levels is probably influenced by the loss of suitable grassland breeding habitat.

The decline has been significantly more severe in Northern Ohio, where winter weather is more extreme, than in Southern Ohio (-14.5 vs. -6.8% annual change, P = 0.004).

CONTINENTAL AND GREAT LAKES REGIONAL TREND: Similarly, Northern Bobwhite have decreased significantly in the region and continent-wide at 2.6 and 2.4% annually.

KILLDEER

Charadrius vociferus

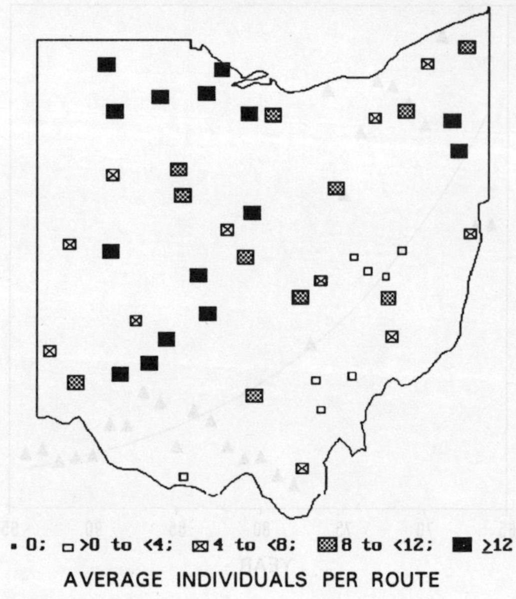

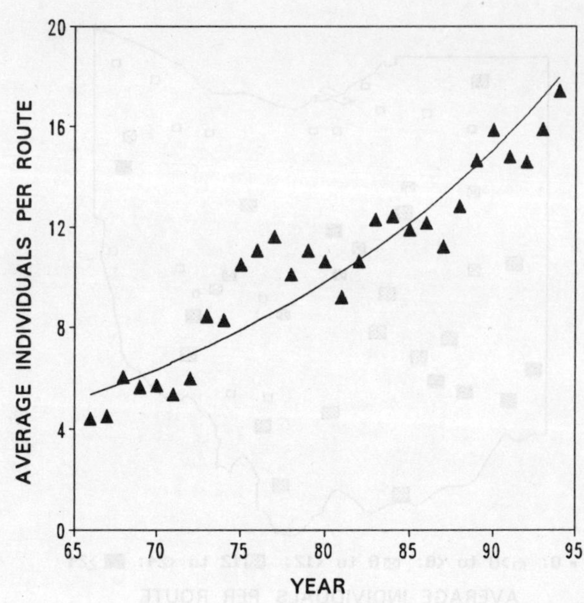

MIGRATORY STATUS: Temperate migrant.

BREEDING HABITAT: Short grasslands and open ground or stone-covered areas.

ABUNDANCE AND DISTRIBUTION: Common (10.5 birds per route) and widely distributed (all routes). Much more common in the predominantly grassland habitat of Western Ohio than in the predominantly forested habitat of Eastern Ohio (14.0 vs 6.5 birds per route, P < 0.001).

OHIO POPULATION TREND: Percent Annual Change (± SE) = 4.4 (± 0.5), P < 0.001
   The 4.4% annual increase in Ohio's Killdeer population was highly significant and has been fairly consistent since 1966. Killdeer increased significantly in both Western and Eastern Ohio, and the two trends were not significantly different from each other (4.9 vs. 3.2%, P = 0.14). Their population increase and widespread distribution apparently results from their ability to thrive in pastures, fields, urban areas, and other disturbed habitats (Peterjohn, 1989).

CONTINENTAL AND GREAT LAKES REGIONAL TREND: Like Ohio's, the regional population also increased significantly at 2.2% annually (P < 0.001), but the continental population has declined significantly since 1966 at 0.5%.

24

SPOTTED SANDPIPER

Actitis macularia

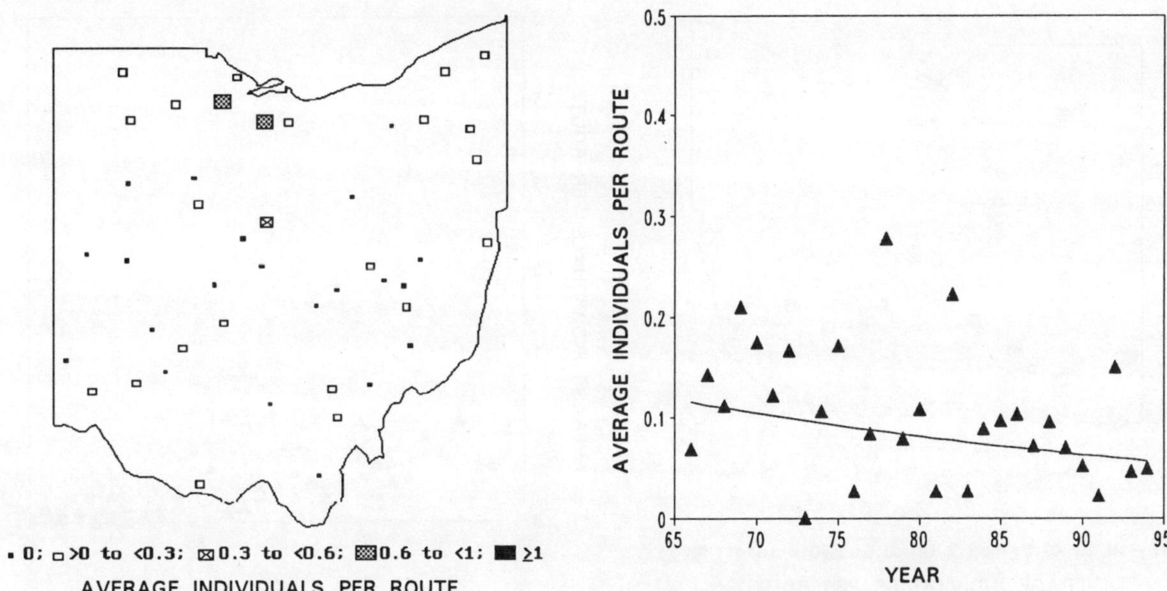

MIGRATORY STATUS: Central and southern neotropical migrant.

BREEDING HABITAT: Wetlands, including lakes, rivers, and streams, especially those with fairly open shorelines.

ABUNDANCE AND DISTRIBUTION: Rare (0.10 birds per route) and locally distributed (24 routes). Spotted Sandpipers do not differ significantly in abundance between Western and Eastern Ohio (0.12 vs. 0.06 birds per route, P = 0.19).

OHIO POPULATION TREND: Percent Annual Change (± SE) = -2.5 (± 2.6), P = 0.35
   Ohio's Spotted Sandpiper population exhibited large annual variation but not a statistically significant overall trend (-2.5% annual change). Trends in Western and Eastern Ohio were similar (-2.9 vs. -2.7%, P = 0.96). Local declines along some lakes and large rivers are probably due to an increase in recreational activities (Peterjohn, 1989; and Peterjohn and Rice, 1991).

CONTINENTAL AND GREAT LAKES REGIONAL TREND: The population decline in the region was substantial (-4.7% annually), showed less annual variation than the Ohio trend, and was highly significant (P < 0.001). The continental population did not exibit a significant trend (-0.2% annually).

UPLAND SANDPIPER

Bartramia longicauda

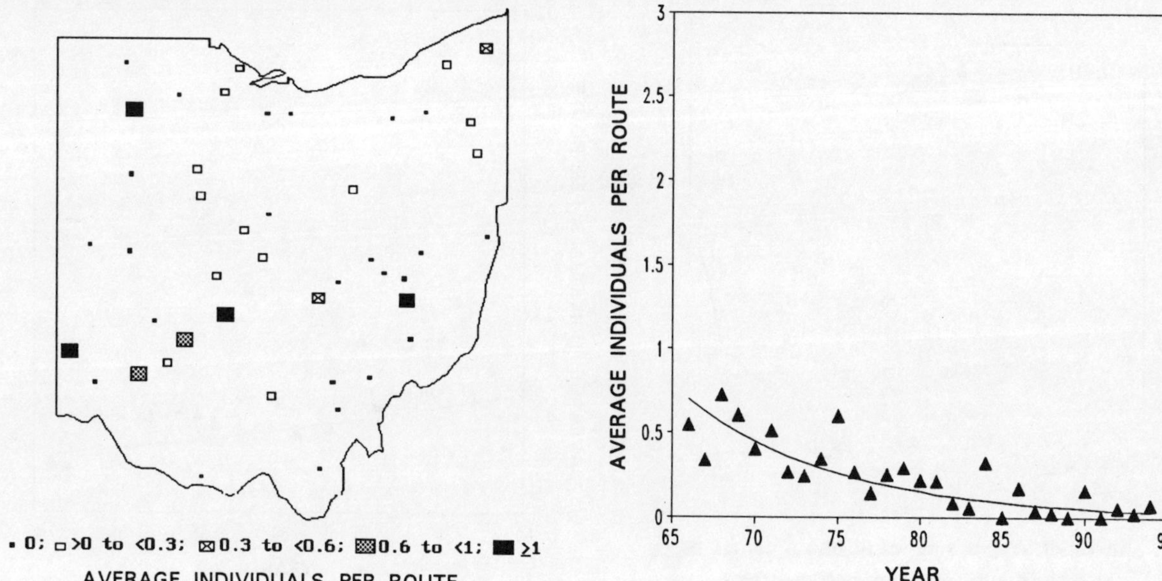

MIGRATORY STATUS: Southern neotropical migrant.

BREEDING HABITAT: Grasslands, particularly open and undisturbed grasslands where vegetation is 1-2 feet high.

ABUNDANCE AND DISTRIBUTION: Rare (0.23 birds per route) and locally distributed (21 routes). Abundance in Western and Eastern Ohio not statistically different (0.28 vs. 0.19 birds per route, P = 0.60).

OHIO POPULATION TREND: Percent Annual Change (± SE) = -10.5 (± 3.0), P = 0.003
Upland Sandpipers have declined significantly at 10.5% annually since 1966. The declining trends were similar in Western and Eastern Ohio (9.2 vs. 12.3% annually, P = 0.51). The decline in Upland Sandpipers is thought to have begun in the 1940s as grasslands were converted to crops and hayfields were mowed more regularly (Peterjohn and Rice, 1991).

CONTINENTAL AND GREAT LAKES REGIONAL TREND: In contrast to Ohio's population, the continental population increased significantly at 2.0% annually and the regional population did not exhibit a significant trend (-0.6%).

ROCK DOVE

Columba livia

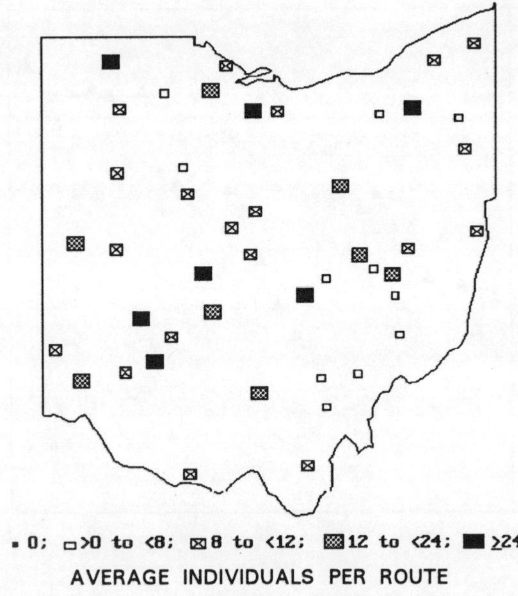

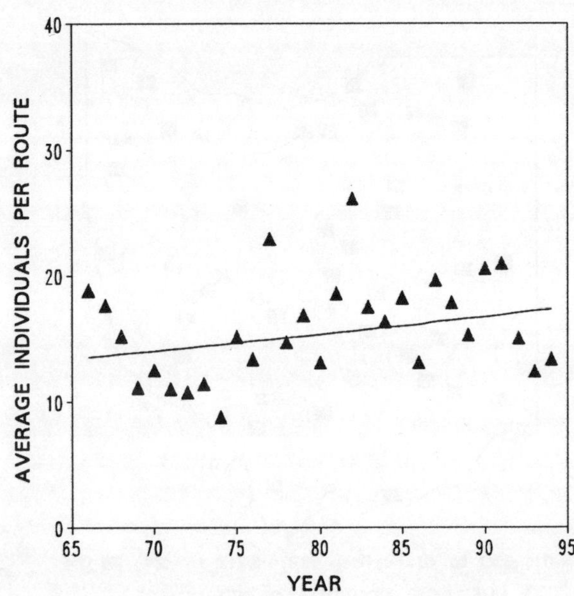

MIGRATORY STATUS: Permanent resident.

BREEDING HABITAT: Usually on human-made structures in rural and urban areas.

ABUNDANCE AND DISTRIBUTION: Abundant (16.0 birds per route) and widely distributed (all routes). Somewhat more abundant in Western than Eastern Ohio, but the difference was not statistically significant (18.8 vs. 12.9, P = 0.11). Probably most numerous in cities. Rock doves were present in Ohio cities during the 1800s and were widespread throughout the state by the 1930s (Peterjohn, 1989).

OHIO POPULATION TREND: Percent Annual Change ($\pm$ SE) = 0.9 ($\pm$ 0.7), P = 0.19
   The overall population change was not significant and exhibited substantial annual variation. However, numbers decreased significantly between 1966 and 1974 at 8.1% annually (P = 0.02), rebounded to earlier levels, and have remained relatively stable since 1975 (-0.3% annual change, P = 0.73). Rock Doves increased significantly at 2.2% annually in Eastern Ohio (P = 0.02) but remained relatively stable in Western Ohio (-0.1%, P = 0.95).

CONTINENTAL AND GREAT LAKES REGIONAL TREND: Neither the continental nor regional population exhibited a significant overall trend (0.4% and -0.3% annual change).

MOURNING DOVE

Zenaida macroura

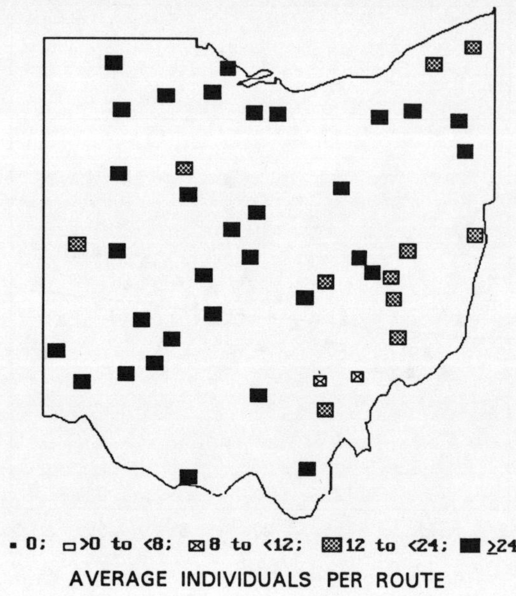

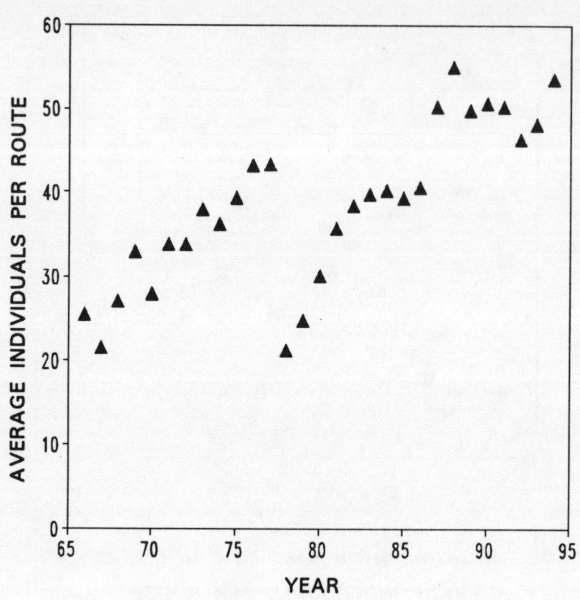

MIGRATORY STATUS: Permanent resident.

BREEDING HABITAT: Residential areas, farmlands, woodlot edges.

ABUNDANCE AND DISTRIBUTION: Abundant (38.2 birds per route) and widely distributed (all routes). More abundant in Western than Eastern Ohio (47.0 vs. 28.3 birds per route, P < 0.001).

OHIO POPULATION TREND: Percent Annual Change calculated for 1966-1994 not appropriate.
   Mourning Doves in Ohio increased steadily through 1977 at 5.8% annually (P < 0.001), declined sharply after the blizzard of 1978, and then increased again at 4.8% annually from 1978-1994 (P < 0.001).
   The population increase exhibited through 1977 is thought to be the continuation of a trend begun during the 19th century as Mourning Doves expanded their range throughout Ohio, perhaps partly as a response to the conversion of forests to agricultural and residential areas (Peterjohn and Rice 1990).
   The pattern of increase, sharp decline, and subsequent increase was similar in Western and Eastern regions of Ohio.

CONTINENTAL AND GREAT LAKES REGIONAL TREND: Neither the continental nor regional population exhibited a significant overall trend (-0.2 and -0.4% annual change). Note that the Mourning Dove's continental and regional trends may be misleading if the extreme winter effect observed in Ohio was evident elsewhere.

BLACK-BILLED CUCKOO

Coccyzus erythropthalmus

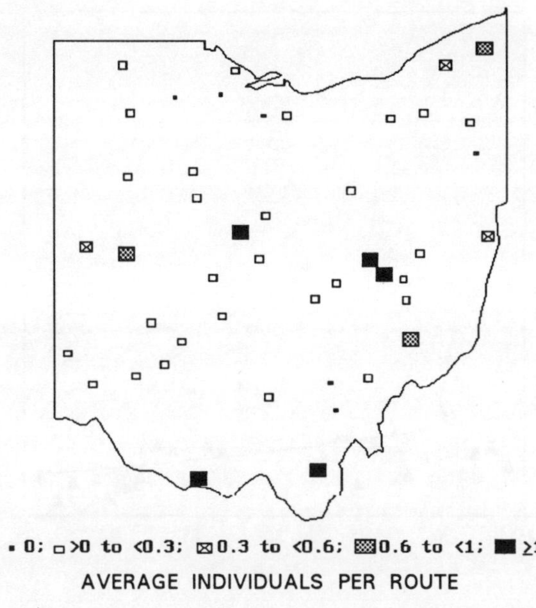

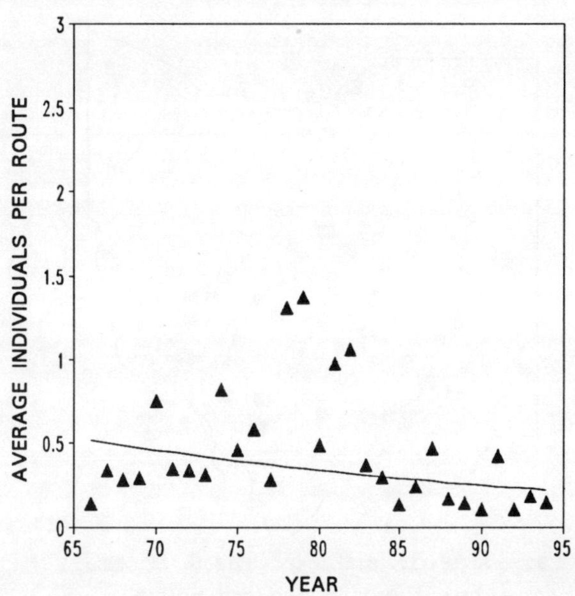

MIGRATORY STATUS: Southern neotropical migrant.

BREEDING HABITAT: Woodlands, especially young woods, edges, and corridors (Peterjohn, 1989).

ABUNDANCE AND DISTRIBUTION: Rare (0.45 birds per route) and fairly widely distributed (39 routes). Equally rare in Western and Eastern Ohio (0.34 vs. 0.56 birds per route, P = 0.29).

OHIO POPULATION TREND: Percent Annual Change (± SE) = -3.0 (±0.9), P = 0.002

The Ohio population has exhibited large annual variation, but has declined significantly at 3.0% annually since 1966. The declines in Western and Eastern Ohio did not differ significantly (4.0 vs. 2.0%, P = 0.31).

The apparent population increase in 1978 and 1979 was exhibited by both Black-billed and Yellow-billed Cuckoos but remains unexplained.

CONTINENTAL AND GREAT LAKES REGIONAL TREND: The continental and regional population both declined at 1.1% annually, but only the continental trend was statistically significant.

YELLOW-BILLED CUCKOO

Coccyzus americanus

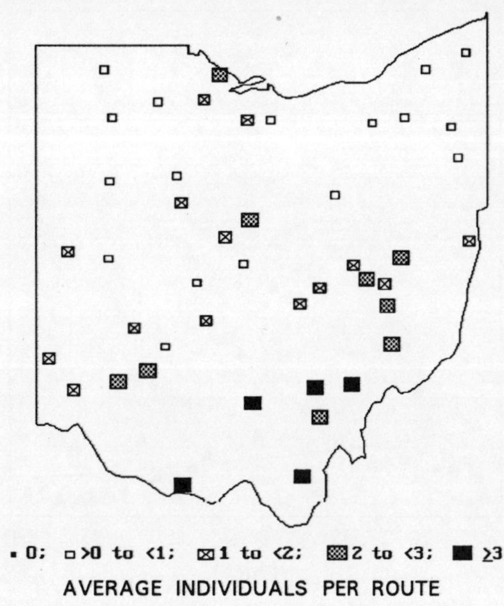

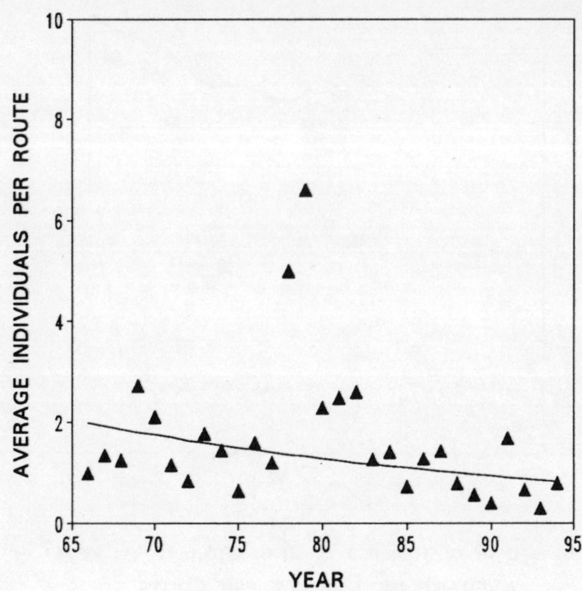

MIGRATORY STATUS: Southern neotropical migrant.

BREEDING HABITAT: Mature or young woodlands.

ABUNDANCE AND DISTRIBUTION: Uncommon (1.6 birds per route) and widely distributed (all routes).
Equally uncommon in Western and Eastern Ohio (1.4 vs. 1.7 birds per route, P = 0.42).

OHIO POPULATION TREND: Percent Annual Change ($\pm$ SE) = -3.1 ($\pm$ 0.6), P < 0.001
   Yellow-billed Cuckoos declined at a significant annual rate of 3.1%, but like Black-billed Cuckoos, they
exhibited considerable annual variation. Population trends for Western and Eastern Ohio were not significantly
different (-3.4 vs. -2.8%, P = 0.50). Both Yellow-billed and Black-billed Cuckoos were observed in high
numbers in 1978 and 1979, an outbreak which remains unexplained. The population trend remains significant
when the 1978 and 1979 outliers are excluded.

CONTINENTAL AND GREAT LAKES REGIONAL TREND: The continental and regional populations
also declined significantly at 1.3% and 2.3%.

GREAT-HORNED OWL

Bubo virginianus

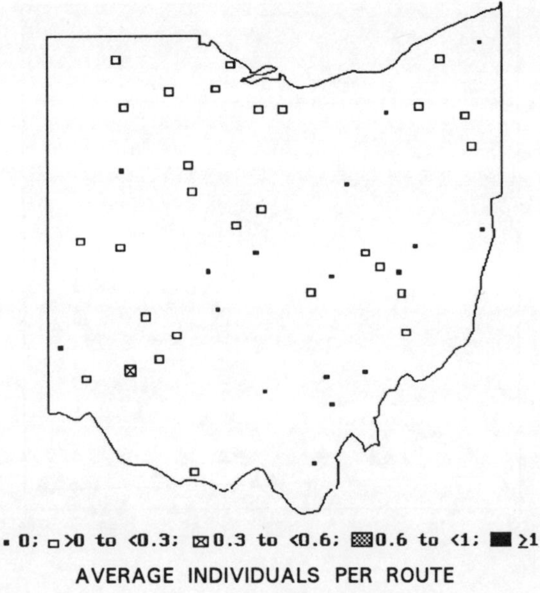

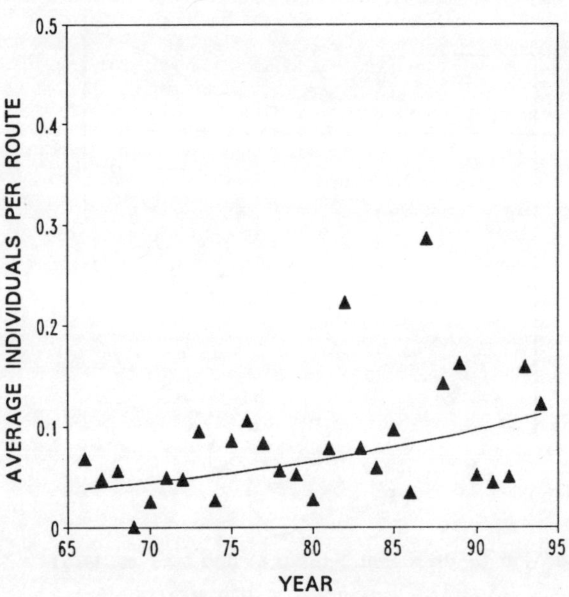

MIGRATORY STATUS: Permanent resident.

BREEDING HABITAT: Mature woodlots interspersed with fields.

ABUNDANCE AND DISTRIBUTION: Rarely recorded (0.08 birds per route) and locally distributed (28 routes). Nocturnal species, such as the Great Horned Owl, are inactive when BBS routes are run and are thus probably more common, relative to other species, than indicated by BBS data.

Great Horned Owls were nearly 3 times more common in Western than Eastern Ohio (0.12 vs. 0.04 birds per route, P = 0.007).

OHIO POPULATION TREND: Percent Annual Change ($\pm$ SE) = 3.9 ($\pm$ 1.6), P = 0.02

The inefficiency of BBS in sampling Great Horned Owls is reflected in the large annual variation in number recorded per route. Although the annual change of 3.9%, as calculated by exponential regression, is statistically significant, the trend should be interpreted with caution. The apparent outliers in 1982 and 1987 are due to high counts on route 61 in 1982 and route 59 in 1987 (each with 3 birds per route, respectively). When the two aberrant counts are removed, the trend remains significant (3.8%, P = 0.02).

In Eastern Ohio, where Great-horned Owls are less common, the population increased significantly at 6.4% annually (P < 0.001); it increased at a lesser, and not quite significant, rate of 2.7% in Western Ohio (P = 0.11).

CONTINENTAL AND GREAT LAKES REGIONAL TREND: Neither the continental nor regional population exhibited a significant trend (1.3 and 2.2% annual change).

BARRED OWL

Strix varia

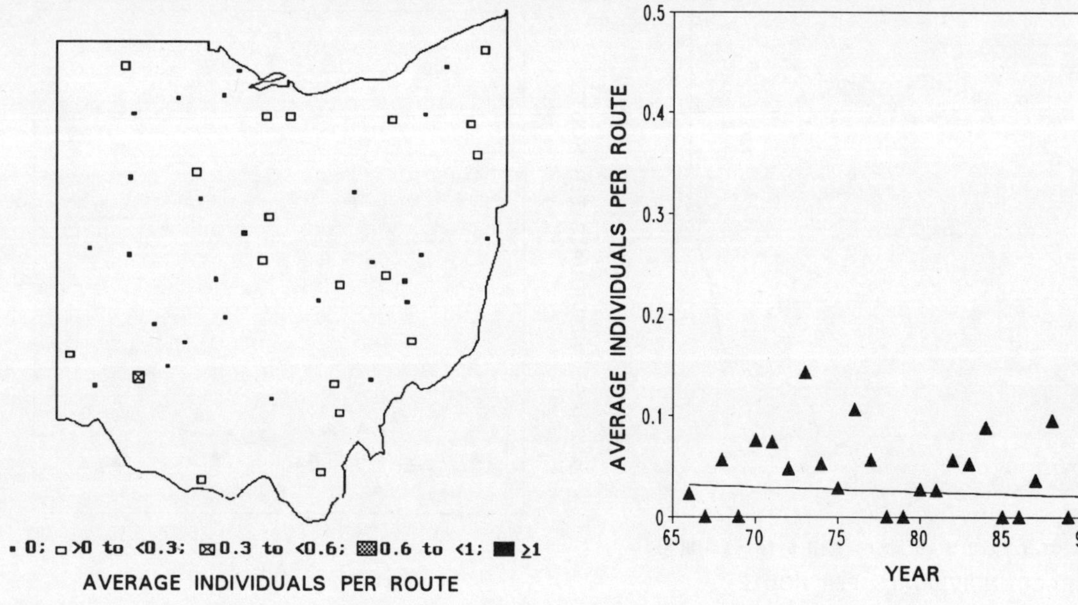

MIGRATORY STATUS: Permanent resident.

BREEDING HABITAT: Extensive mature woodlands with large trees and cavities suitable for nesting.

ABUNDANCE AND DISTRIBUTION: Rarely recorded (0.04 birds per route) and very locally distributed (19 routes). Like other nocturnal species, Barred Owls are underrepresented in BBS counts.

Barred Owl abundance did not differ on Western and Eastern BBS routes (0.04 vs. 0.04 birds per route, P = 0.93), but they occurred less frequently in Western than Eastern Atlas blocks (33 vs. 75%; Peterjohn and Rice, 1991).

Their scarcity in the Western Region is believed to have begun in the early 1800s when virgin forests were cleared and to continue today due to intensive farming (Peterjohn, 1989). Barred Owls require extensive forests and are outcompeted by Great Horned Owls in small, fragmented woodlands like those found in western Ohio (Peterjohn, 1989).

OHIO POPULATION TREND: Percent Annual Change ($\pm$ SE) = -1.8 ($\pm$ 2.2), P = 0.44

Like Great Horned Owls, Barred Owls exhibit substantial annual variation that probably results from the poor sampling efficiency of BBS. The overall population trend of -1.8% annual change is not statistically significant and should be interpreted with caution because it is greatly influenced by the low counts in 1989-1991. The sample was inadequate to compare Western and Eastern population trends.

CONTINENTAL AND GREAT LAKES REGIONAL TREND: Both continental and regional populations increased (4.6 and 11.6% annually), but only the continental trend was significant.

COMMON NIGHTHAWK

Chordeiles minor

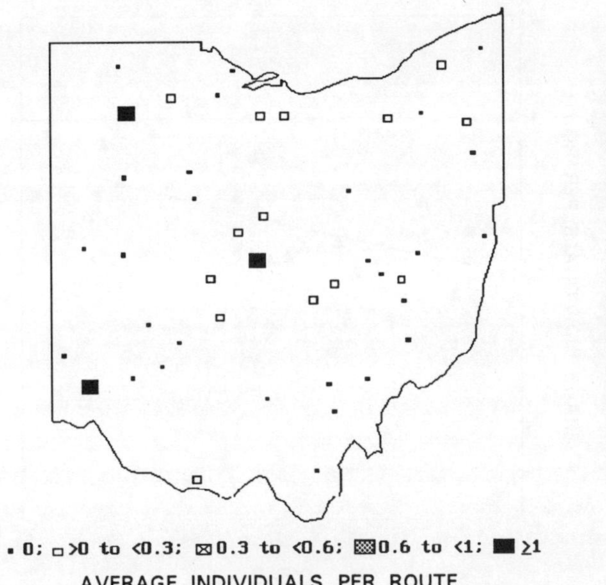

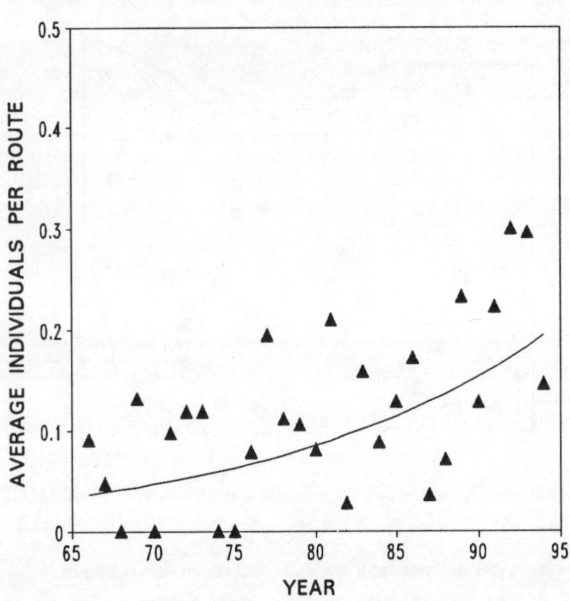

MIGRATORY STATUS: Southern neotropical migrant.

BREEDING HABITAT: Primarily on flat rooftops in residential and urban areas.

ABUNDANCE AND DISTRIBUTION: Rarely recorded (0.11 birds per route) and very locally distributed (17 routes). The abundance of the Common Nighthawk, like that of other nocturnal species, is underestimated, relative to other species, by the BBS technique.

Somewhat more abundant in Western than Eastern Ohio, but the difference was not statistically significantly (0.19 vs. 0.03 birds per route, P = 0.07). Associated primarily with urban areas today and throughout the 1900s (Peterjohn, 1989).

OHIO POPULATION TREND: Percent Annual Change (± SE) = 6.1 (± 2.9), P = 0.04

The inefficiency of BBS in sampling the Common Nighthawk population was reflected in the large annual variation in number recorded per route. Although the overall increase of 6.1% annually was statistically significant, it should be interpreted with caution. Despite the growth of their favored habitat, urban areas, knowledgeable observers believe that Common Nighthawks have declined in recent years (Peterjohn, 1989).

The sample was inadequate to compare Western and Eastern population trends.

CONTINENTAL AND GREAT LAKES REGIONAL TREND: In contrast to the Ohio population, the continental population of Common Nighthawks has declined significantly at 0.9% annually since 1966. The regional population did not exhibit a significant annual change (0.3%).

CHIMNEY SWIFT

Chaetura pelagica

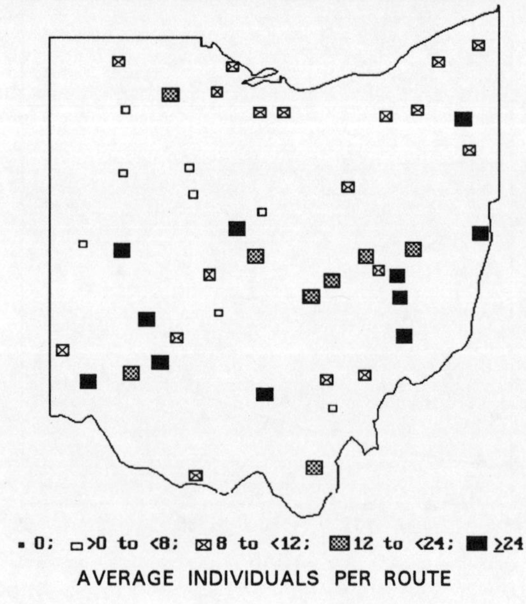

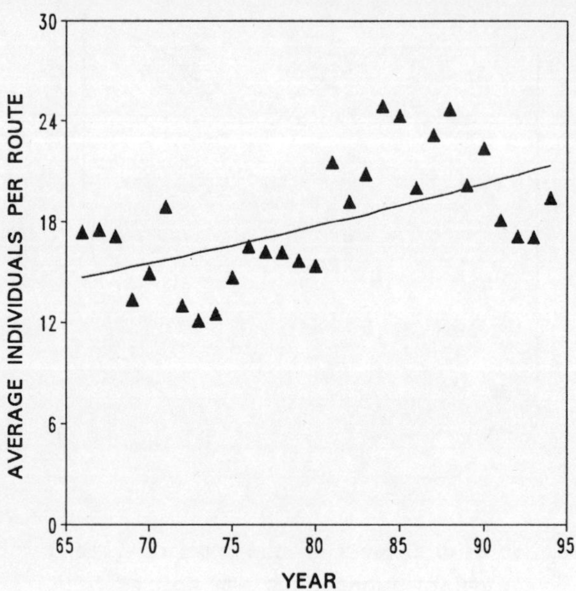

MIGRATORY STATUS: Southern neotropical migrant.

BREEDING HABITAT: Residential areas.

ABUNDANCE AND DISTRIBUTION: Abundant (18.2 birds per route) and widely distributed (all routes). Approximately equally abundant in Western and Eastern Ohio (16.1 vs. 20.6 birds per route, P = 0.19).

OHIO POPULATION TREND: **Percent Annual Change ($\pm$ SE) = 1.3 ($\pm$ 0.7), P = 0.046**
   Chimney swifts have increased significantly at 1.3% annually since 1966. The trend in BBS data is obvious, despite large annual variation in the number recorded. The trend in Eastern Ohio was significantly increasing (2.5%, P = 0.03), the trend in Western Ohio was highly variable (0.2%, P = 0.72), and the two trends were almost significantly different from one another (P = 0.07).
   Ohio's Chimney Swift population probably began its increase during the 19th and early 20th centuries as the number of buildings and chimneys increased and the nesting behavior of the birds changed to take advantage of the new habitat (Peterjohn and Rice, 1991).

CONTINENTAL AND GREAT LAKES REGIONAL TREND: In contrast to Ohio's population, the regional and continental populations both exhibited significant declines of 1.0% annually.

RUBY-THROATED HUMMINGBIRD

Archilochus colubris

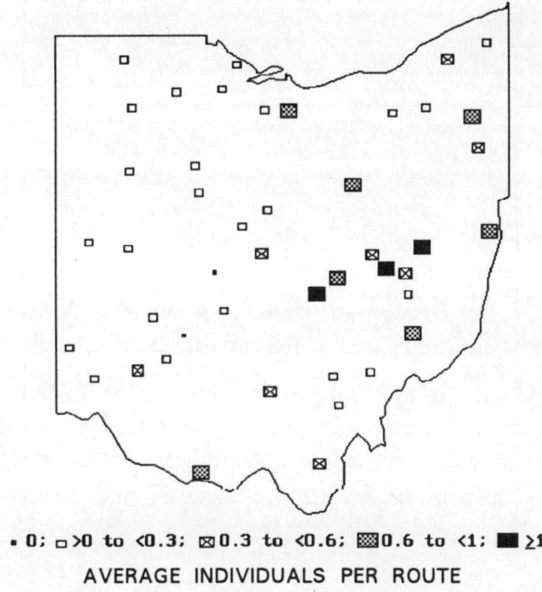

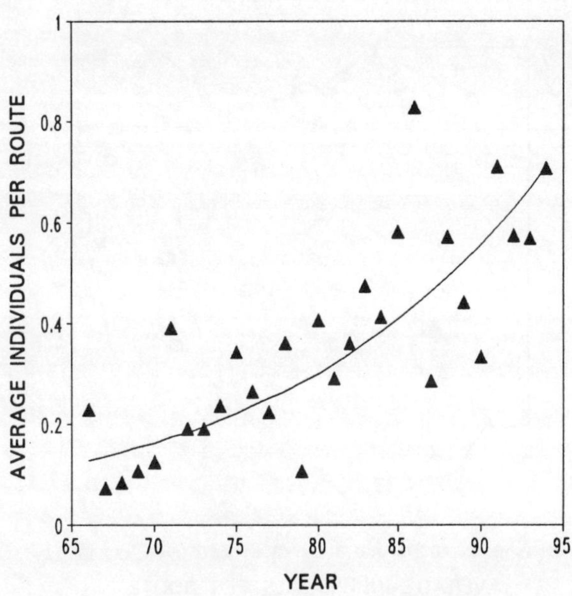

MIGRATORY STATUS: Central neotropical migrant.

BREEDING HABITAT: Mature and second-growth woodlands.

ABUNDANCE AND DISTRIBUTION: Rarely recorded (0.36 birds per route) but widely distributed (43 routes). Because their vocalizations are difficult to detect relative to those of most species, Ruby-throated Hummingbirds are probably relatively more common in Ohio than suggested by BBS data.
   More common in Eastern than Western Ohio (0.55 vs. 0.20 birds per route, P < 0.001).

OHIO POPULATION TREND: Percent Annual Change (± SE) = 6.9 (± 1.0), P < 0.001
   Ruby-throated Hummingbirds increased significantly at 6.9% annually. The trends in Western and Eastern Ohio were similar (5.8 vs. 7.8% annually, P = 0.42).

CONTINENTAL AND GREAT LAKES REGIONAL TREND: The continental population also increased significantly at 1.1% annually. The regional population did not exhibit a significant trend (0.2%).

BELTED KINGFISHER

Ceryle alcyon

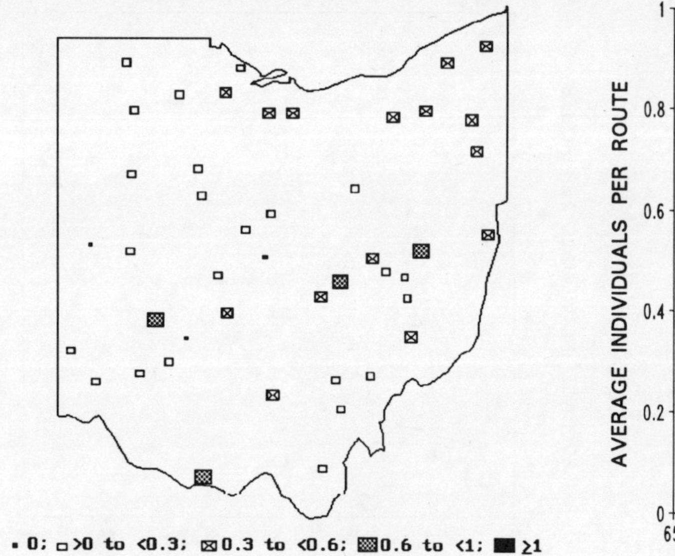

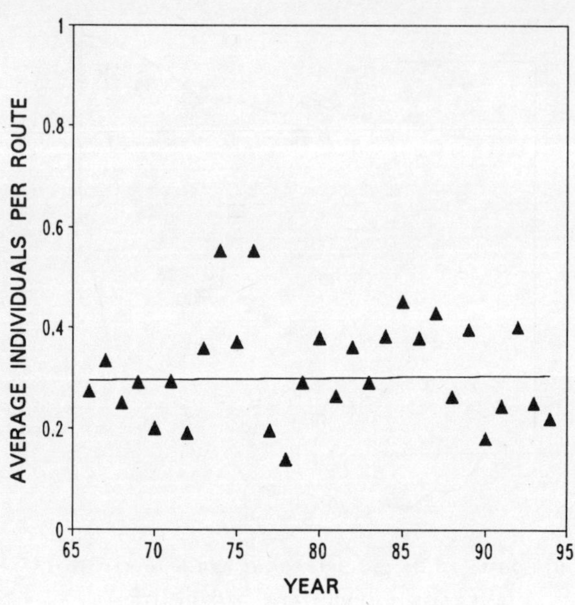

MIGRATORY STATUS: Temperate migrant, although some may winter in Ohio in locations with open water.

BREEDING HABITAT: Streams, lakes, and ponds.

ABUNDANCE AND DISTRIBUTION: Rarely recorded (0.31 birds per route) but widely distributed (42 routes). Significantly more common in Eastern than Western Ohio (0.39 vs. 0.24 birds per route, P = 0.03).

OHIO POPULATION TREND: **Percent Annual Change (± SE) = 0.1 (± 0.9), P = 0.89**

Ohio's Belted Kingfisher population has not exhibited a significantly increasing or decreasing trend since 1966 (0.1% annual change). Trends for Eastern and Western Ohio were similar (0.03 vs. -0.06%, P =0.96).

Historic records indicate that the status of Belted Kingfishers has not changed markedly during this century, but have declined in some watersheds due to channelization or deterioration in water quality (Peterjohn and Rice, 1991).

CONTINENTAL AND GREAT LAKES REGIONAL TREND: As in Ohio, Belted Kingfishers in the region have not exhibited a significant trend (0.1%), however, the continental population declined significantly at 1.9% annually.

RED-HEADED WOODPECKER

Melanerpes erythrocephalus

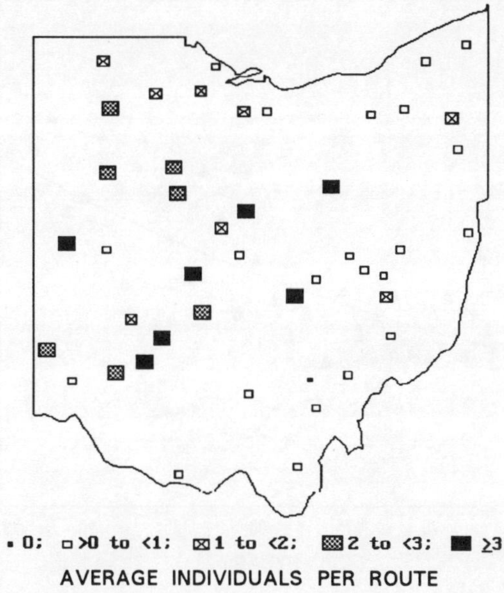

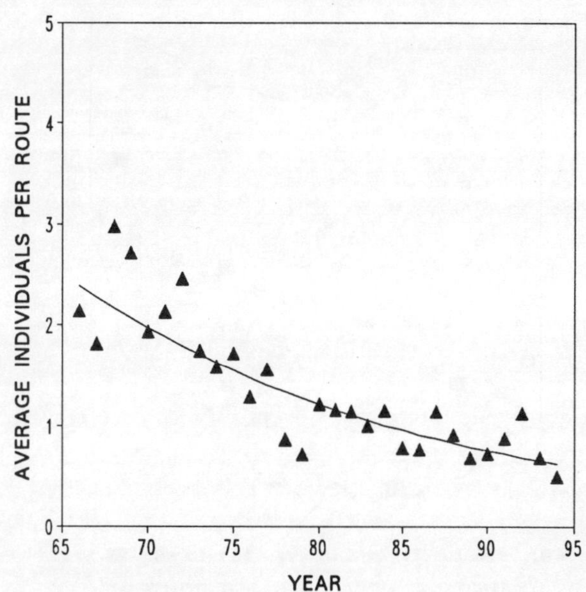

MIGRATORY STATUS: Temperate migrant, but many winter in Ohio during years with abundant mast.

BREEDING HABITAT: Mature and young woodlands.

ABUNDANCE AND DISTRIBUTION: Uncommon (1.4 birds per route) but widely distributed (44 routes).
More common in Western than Eastern Ohio (2.0 vs. 0.8 birds per route, P = 0.002). Presumably they prefer
the rather open woodlands found in Western Ohio to the mature forests in the east.

OHIO POPULATION TREND: **Percent Annual Change (+ SE) = -4.7 (+ 0.8), P < 0.001**
   Red-headed Woodpeckers declined severely at 4.7% annually during 1966-1994. Declines in Western and
Eastern Ohio were similar (4.8 vs. 5.0% annually, P = 0.94).
   Historical records indicate that Red-headed Woodpeckers probably increased during the late 1800s as dense
forests were cleared and fields became interspersed with woodlands (Peterjohn, 1989). However, a decline in
Red-headed Woodpeckers has been evident throughout the 20th century; it is attributed to the spread of
European Starlings in the early decades, and more recently, to cutting of suitable woodlands in the west, which
supports the majority of Ohio's population (Peterjohn and Rice, 1991).

CONTINENTAL AND GREAT LAKES REGIONAL TREND: The continental and regional
populations also declined significantly at 1.9 and 3.0% annually.

RED-BELLIED WOODPECKER

Melanerpes carolinus

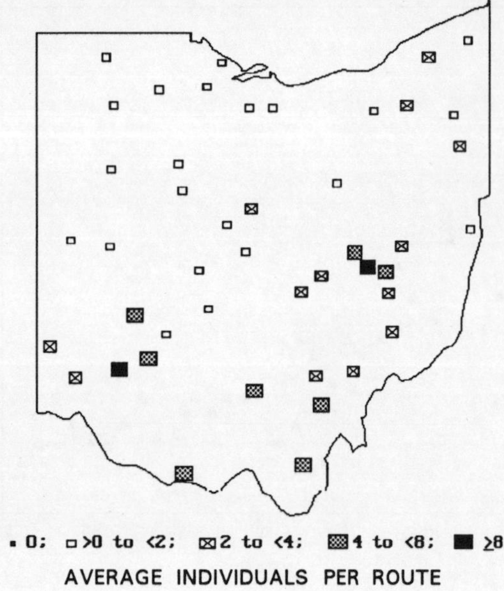

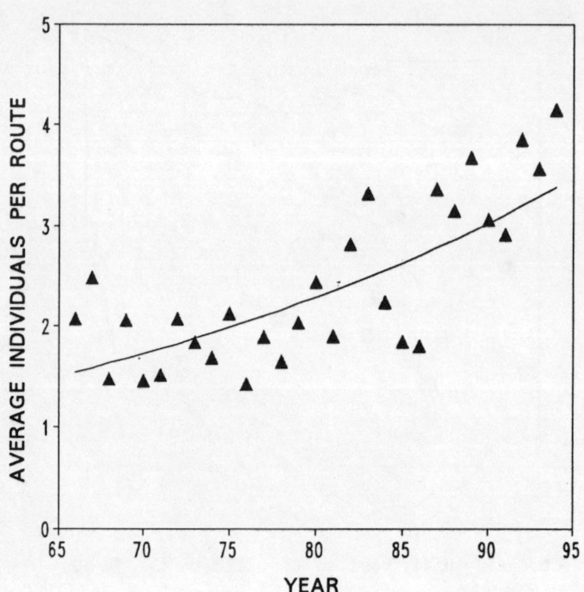

MIGRATORY STATUS: Permanent resident.

BREEDING HABITAT: Mature and young woodlands, including small woodlots and wooded residential areas.

ABUNDANCE AND DISTRIBUTION: Fairly common (2.4 birds per route) and widely distributed (all routes). Twice as abundant in Eastern than Western Ohio (3.3 vs. 1.6 birds per route, P = 0.003). Typically more common in the southern U.S., Red-bellied Woodpeckers were also more abundant in Southern than Northern Ohio (3.3 vs. 1.1 birds per route, P < 0.001).

OHIO POPULATION TREND: Percent Annual Change (± SE) = 2.9 (± 0.6), P < 0.001
   Red-bellied Woodpeckers increased significantly at 2.9% annually. Trends for Eastern and Western Ohio did not differ significantly (2.5 vs. 3.4% annually, P = 0.45), however Red-bellied Woodpeckers increased more rapidly in Northern than Southern Ohio (5.4 vs. 2.0%, P = 0.03). The increase may reflect their northward range expansion throughout eastern North America; the expansion was noticeable in the northern third of Ohio beginning in the 1940s or 1950s (Peterjohn, 1989).

CONTINENTAL AND GREAT LAKES REGIONAL TREND: The continental and regional populations also increased significantly at 0.6 and 1.3% annually.

DOWNY WOODPECKER

Picoides pubescens

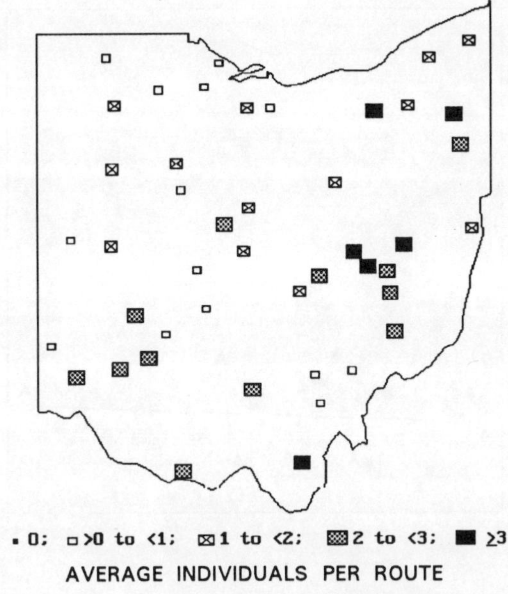

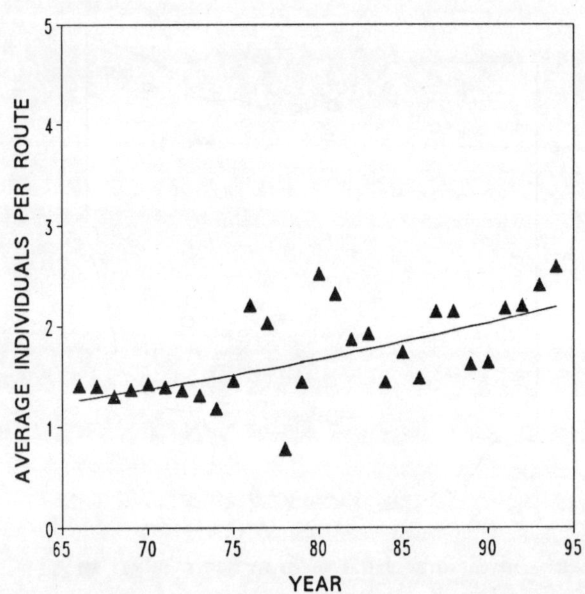

MIGRATORY STATUS: Permanent resident.

BREEDING HABITAT: All types of woodlands.

ABUNDANCE AND DISTRIBUTION: Uncommonly recorded on BBS routes (1.8 birds per route) but are underrepresented, relative to other species, by BBS because they seldom vocalize in early June; widely distributed (all routes). More common in Eastern than Western Ohio (2.2 vs. 1.3 birds per route, P = 0.01).

OHIO POPULATION TREND: Percent Annual Change (± SE) = 2.0 (± 0.8), P = 0.02
Downy Woodpeckers increased significantly at 2.0% annually. The increase was large and significant in Western Ohio (4%, P < 0.001) but not in Eastern Ohio (0.6%, P = 0.58).

CONTINENTAL AND GREAT LAKES REGIONAL TREND: The regional population also increased significantly at 1.0% annually, however, the continental population did not exhibit a significant trend (0.0%).

HAIRY WOODPECKER

Picoides villosus

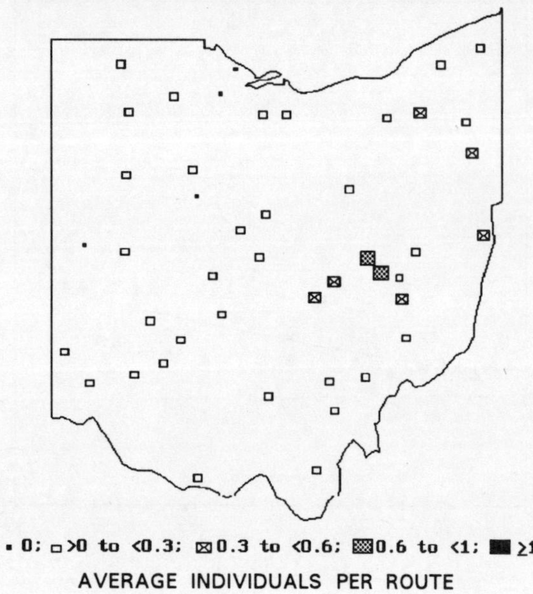

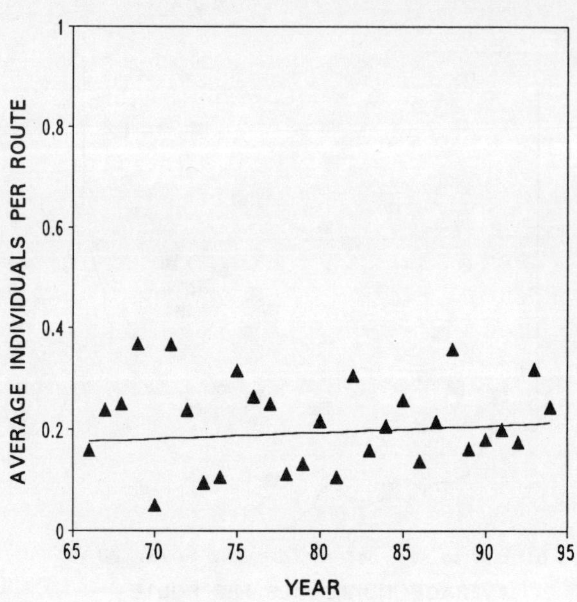

MIGRATORY STATUS: Permanent resident.

BREEDING HABITAT: Mature woodlands.

ABUNDANCE AND DISTRIBUTION: Rarely recorded on BBS routes (0.2 birds per route) but widely distributed (41 routes). Like Downy Woodpeckers, Hairy Woodpeckers do not often vocalize during summer months and are thus probably underrepresented by BBS relative to other species.

More common in Eastern than Western Ohio (0.34 vs. 0.11 birds per route, P < 0.001).

OHIO POPULATION TREND: Percent Annual Change (± SE) = 0.7 (± 1.2), P = 0.60

Ohio's Hairy Woodpecker population did not exhibit a significant trend during 1966-1994 (0.7% annual change). Hairy Woodpeckers tended to increase in Eastern Ohio at 2.0% annually (P = 0.22) and decrease in Western Ohio at 1.9% (P = 0.23), although neither trend was significant.

Historic records also indicate a fairly stable population, although there have been local declines in Western Ohio due to cutting of mature woodlands (Peterjohn, 1989).

CONTINENTAL AND GREAT LAKES REGIONAL TREND: Like the Ohio population, the regional population did not exhibit a significant annual change (0.8%), however, the continental population has increased at 1.2% annually.

NORTHERN FLICKER

Colaptes auratus

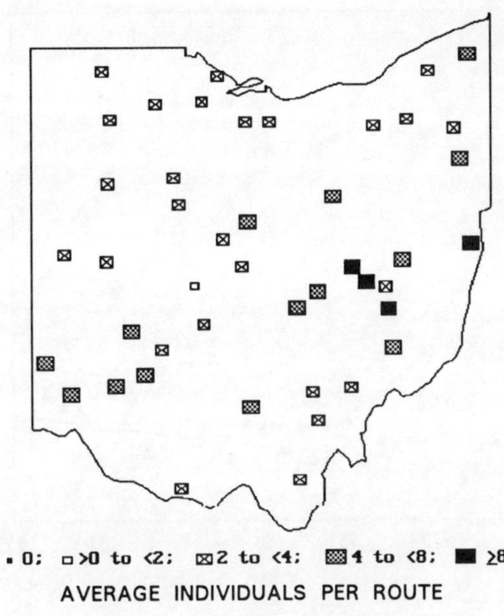

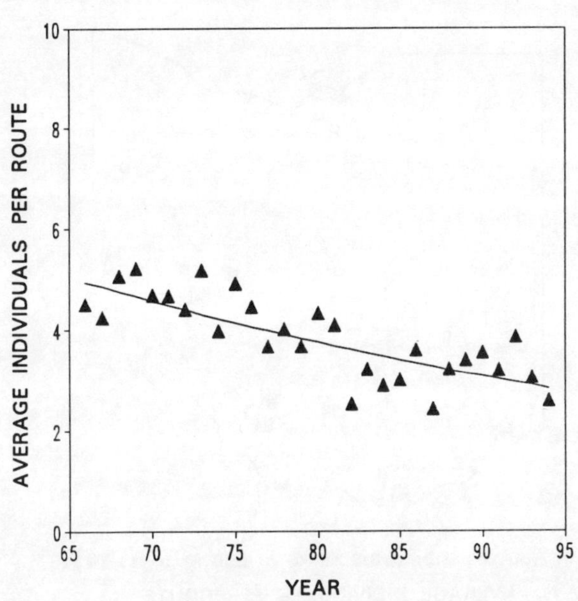

MIGRATORY STATUS: Temperate migrant.

BREEDING HABITAT: Woodlands, especially open woodlots and edges with nearby grassy areas.

ABUNDANCE AND DISTRIBUTION: Fairly common (3.9 birds per route) and widely distributed (all routes). Significantly more common in Eastern than Western Ohio (4.6 vs. 3.3 birds per route, P = 0.004).

OHIO POPULATION TREND: Percent Annual Change ($\pm$ SE) = **-2.0 ($\pm$ 0.4), P < 0.001**
  Northern Flickers declined significantly in Ohio at 2.0% annually. They declined significantly in both Eastern Ohio at 2.0% annually (P = 0.002) and in Western Ohio at 1.9% (P = 0.003).

CONTINENTAL AND GREAT LAKES REGIONAL TREND: The continental and regional populations of Yellow-shafted Flickers (the eastern race of the Northern Flicker) also declined significantly at 2.9 and 3.2% annually. As in Ohio, declines are thought to result, in part, from competition with European Starlings. In addition, the effect of high mortality during the severe winters of 1976-77 and 1977-78 is also evident in the continental population (Robbins et al., 1986).

PILEATED WOODPECKER

Dryocopus pileatus

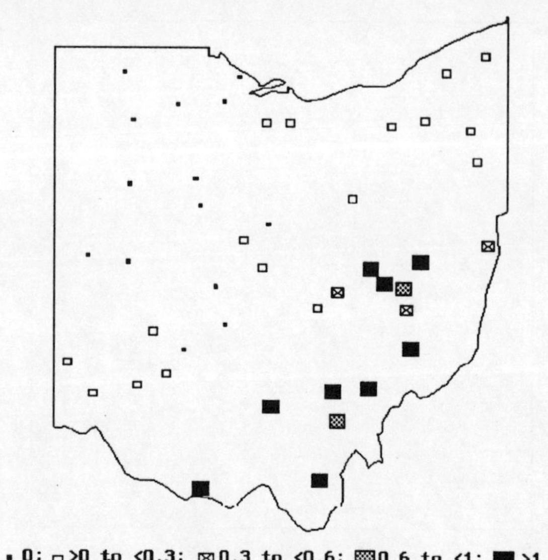

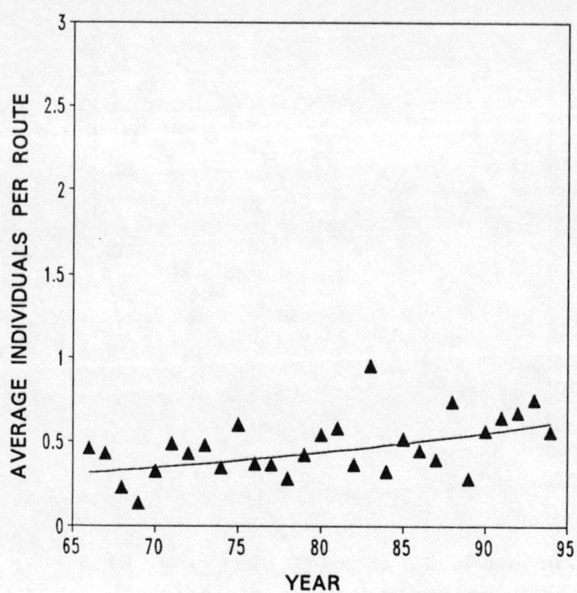

MIGRATORY STATUS: Permanent resident.

BREEDING HABITAT: Mature woodlands, especially large undisturbed tracts.

ABUNDANCE AND DISTRIBUTION: Rare (0.47 birds per route) and fairly widely distributed (31 routes). Pileated Woodpeckers were >6 times more common in Eastern than Western Ohio (0.86 vs. 0.13 birds per route, P < 0.001).

OHIO POPULATION TREND: Percent Annual Change (± SE) = 2.4 (± 0.4), P < 0.001

Pileated Woodpeckers increased significantly in Ohio at 2.4% annually. The trend in Eastern Ohio was significant (2.1%, P <0.001) and the sample in Western Ohio was inadequate to estimate a trend.

As the extensive mature forests of Ohio were cleared during the 1800s, Pileateds were extirpated from many counties. Their recovery began in the 1920s and 1930s as cleared land in Eastern Ohio began to revert to forest (Peterjohn, 1989).

CONTINENTAL AND GREAT LAKES REGIONAL TREND: The continental and regional populations also increased significantly at 2.0 and 3.8% annually.

EASTERN WOOD-PEWEE

Contopus virens

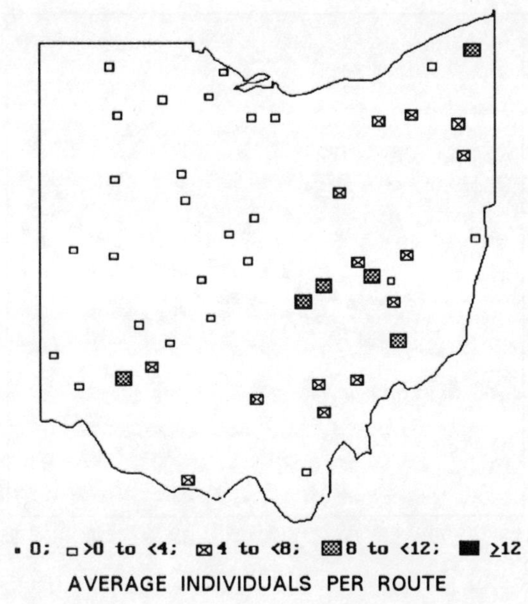

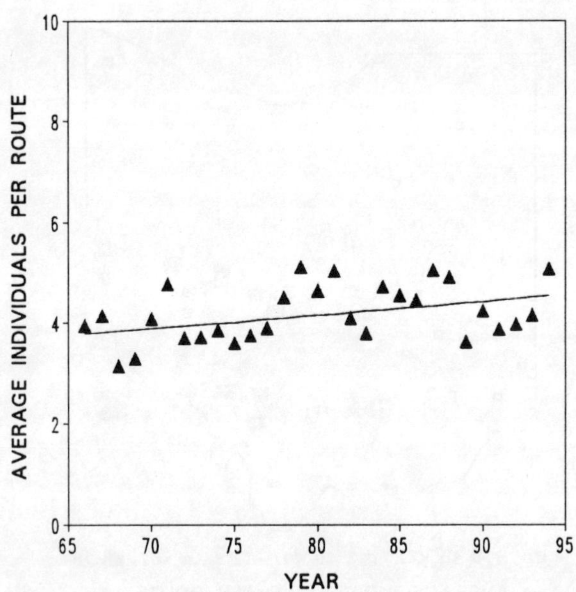

MIGRATORY STATUS: Southern neotropical migrant.

BREEDING HABITAT: Mature woodlands, including small woodlots and woodland edges.

ABUNDANCE AND DISTRIBUTION: Common (4.2 birds per route) and widely distributed (all routes). More than twice as common in Eastern than Western Ohio (6.3 vs. 2.3 birds per route, P < 0.001).

OHIO POPULATION TREND: Percent Annual Change ($\pm$ SE) = 0.7 ($\pm$ 0.5), P = 0.23
  Ohio's population did not exhibit a significant trend during 1966 1994 (0.7% annual change). However, Eastern Wood-Pewees increased at a nearly significant rate of 1.3% annually (P = 0.06) in Western Ohio, where they are less common, but not in Eastern Ohio (0.4%, P = 0.61).

CONTINENTAL AND GREAT LAKES REGIONAL TREND: Both the continental and regional populations have decreased significantly at 1.7 and 0.7% annually.

ACADIAN FLYCATCHER

Empidonax virescens

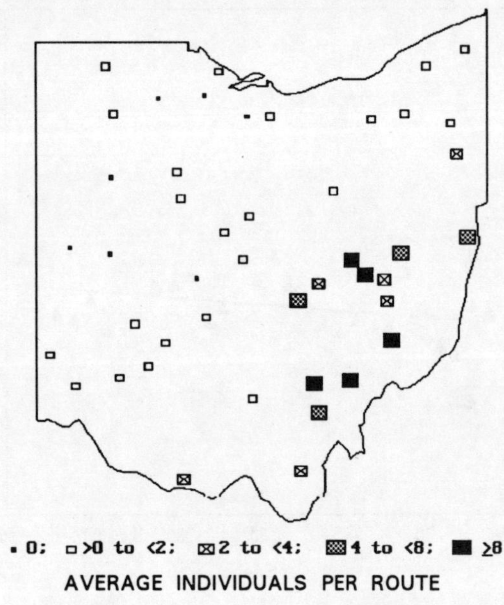

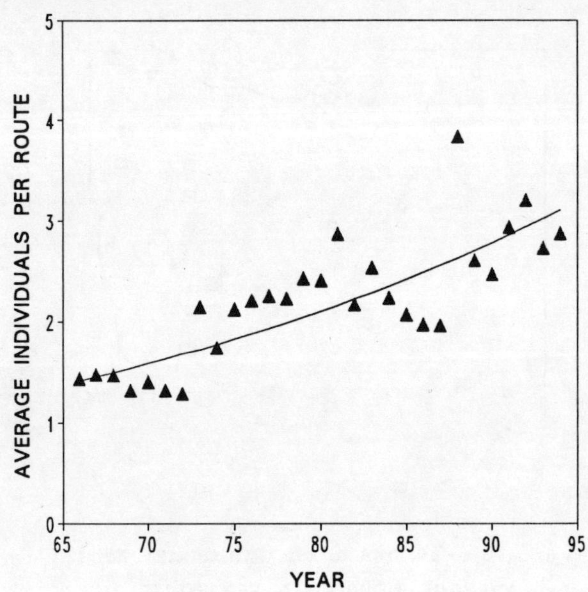

MIGRATORY STATUS: Southern neotropical migrant.

BREEDING HABITAT: Mature woodlands, particularly the interiors of extensive woodlands and often near small streams.

ABUNDANCE AND DISTRIBUTION: Fairly common overall (2.2 birds per route) and fairly widely distributed (38 routes), Acadian Flycatchers are much more common in Eastern than Western Ohio (4.3 vs. 0.3 birds per route, P < 0.001).

OHIO POPULATION TREND: Percent Annual Change (± SE) = 2.8 (± 1.1), P = 0.01
   Overall, Acadian Flycatchers increased significantly at 2.8% annually. The increase was substantial in Eastern Ohio (3.2% annually, P = 0.01) but not in Western Ohio (0.5%, P = 0.83).

CONTINENTAL AND GREAT LAKES REGIONAL TREND: In contrast to Ohio's population of Acadian Flycatchers, the regional population is decreasing significantly (1.6%) and the continental population did not exhibit a significant 1966-1991 trend (0.5% annual change).

WILLOW FLYCATCHER

Empidonax traillii

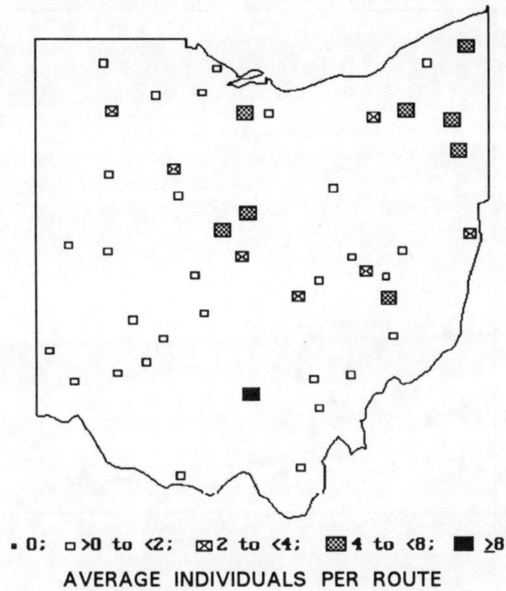

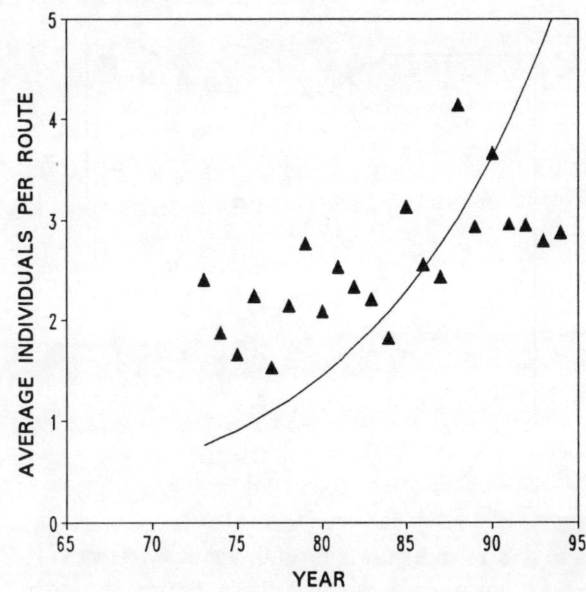

MIGRATORY STATUS: Central and southern neotropical migrant.

BREEDING HABITAT: Scrub, especially bordering wetlands and streams.

ABUNDANCE AND DISTRIBUTION: Fairly common (2.0 birds per route) and widely distributed (45 routes). Abundances in Eastern and Western Ohio do not differ significantly (2.4 vs. 1.6 birds per route, P = 0.14).

OHIO POPULATION TREND: Percent Annual Change ($\pm$ SE) = 2.6 ($\pm$ 1.5), P = 0.09
   Willow and Alder Flycatchers were classified as one species, the Traill's Flycatcher, until 1972. Since 1973, Willow Flycatchers have increased somewhat, but nonsignificantly, at 2.6% annually. Alder Flycatchers are observed too infrequently on BBS routes to allow a reasonably precise estimation of population trend.
   Trends of Willow Flycatchers in Eastern and Western Ohio did not differ significantly (2.8 vs. 2.2%, P = 0.40).

CONTINENTAL AND GREAT LAKES REGIONAL TREND: Regional and continental trends were not calculated because data does not begin until 1972.

LEAST FLYCATCHER

Empidonax minimus

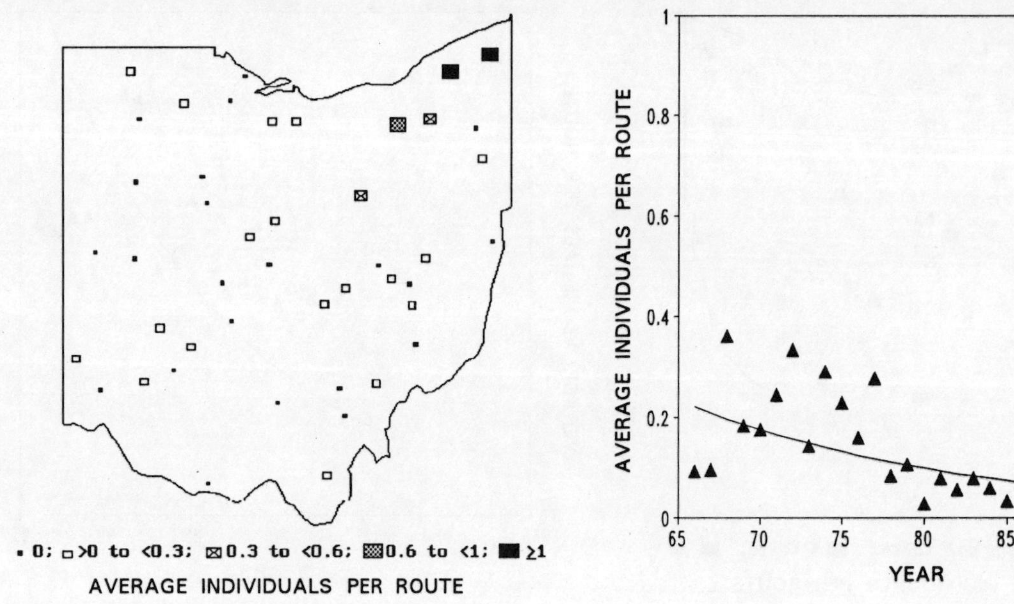

MIGRATORY STATUS: Central neotropical migrant.

BREEDING HABITAT: Young woods, edges, riparian corridors, or openings of disturbed woods.

ABUNDANCE AND DISTRIBUTION: Rare (0.12 birds per route) and locally distributed (23 routes). Ohio is on the southern edge of the Least Flycatcher's breeding range, and they are more than 7 times more common in northern than southern Ohio (0.22 vs. 0.03 birds per route, P = 0.04). It is thought that their breeding range expanded southward beginning in the 1930s. Also more common in the predominantly forested Eastern Ohio than in the farmlands of Western Ohio (0.22 vs. 0.03 birds per route, P = 0.04).

OHIO POPULATION TREND: Percent Annual Change ($\pm$ SE) = -5.6 ($\pm$ 2.1), P = 0.01
   Least Flycatchers have declined significantly in Ohio since 1966 (5.6% annually), but have been relatively stable since 1980 (1.1%, P = 0.46). Although declining significantly at 6.9% annually (P = 0.004) in northern Ohio where they are more common, Least Flycatchers are increasing significantly in Southern Ohio (3.4%, P = 0.03), apparently as a result of southward range expansion.

CONTINENTAL AND GREAT LAKES REGIONAL TREND: Like the Ohio population, the regional and continental populations also declined significantly at 2.3 and 0.8%.

EASTERN PHOEBE

Sayornis phoebe

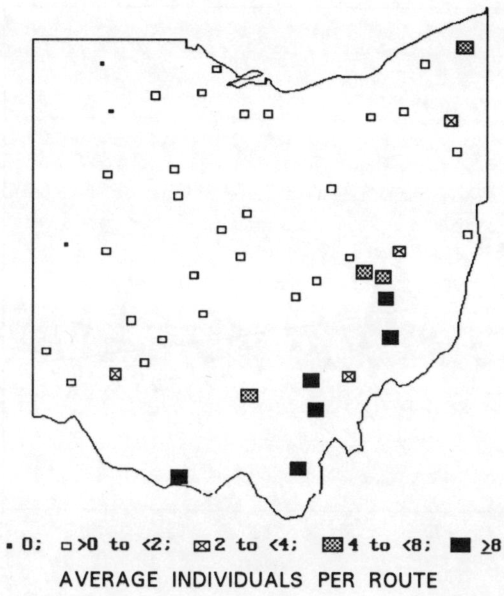

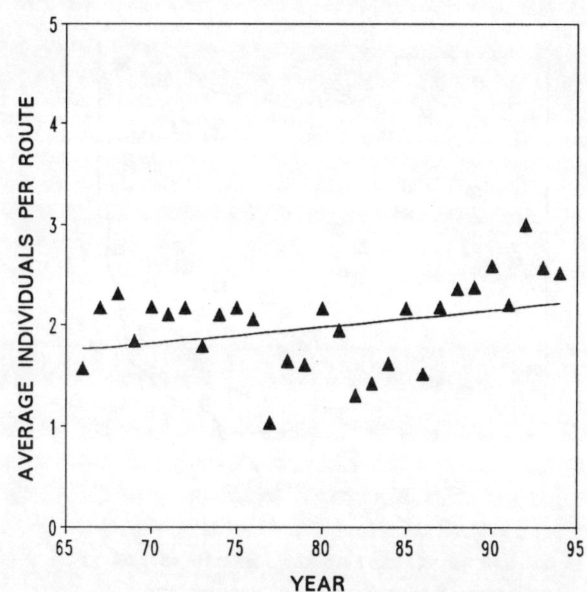

MIGRATORY STATUS: Temperate migrant.

BREEDING HABITAT: Scrub, particularly wooded riparian corridors and other woodland edges.

ABUNDANCE AND DISTRIBUTION: Fairly common (2.0 birds per route) and widely distributed (42 routes). More than 4 times more common in Eastern than Western Ohio (3.4 vs. 0.7 birds per route, P < 0.001), particularly common in the southeastern Unglaciated Plateau.

OHIO POPULATION TREND: Percent Annual Change ($\pm$ SE) = 0.8 ($\pm$ 0.7), P = 0.25

Ohio's population did not exhibit a significant 1966-94 trend (0.8% annual change). Trends in Eastern and Western Ohio did not differ (0.7 vs. 1.4%, P = 0.49).

Historically, numbers of Eastern Phoebes probably increased as early settlers created more edge habitats, but their populations are thought to have declined during the 1940s-60s, particularly in intensively farmed Western Ohio (Peterjohn, 1989). In addition to using human-induced edge habitat, phoebes nest on human-made structures such as bridges, and outbuildings in rural areas.

CONTINENTAL AND GREAT LAKES REGIONAL TREND: The continental and regional populations increased at 0.9% annually, although only the continental trend was significant.

GREAT CRESTED FLYCATCHER

Myiarchus crinitus

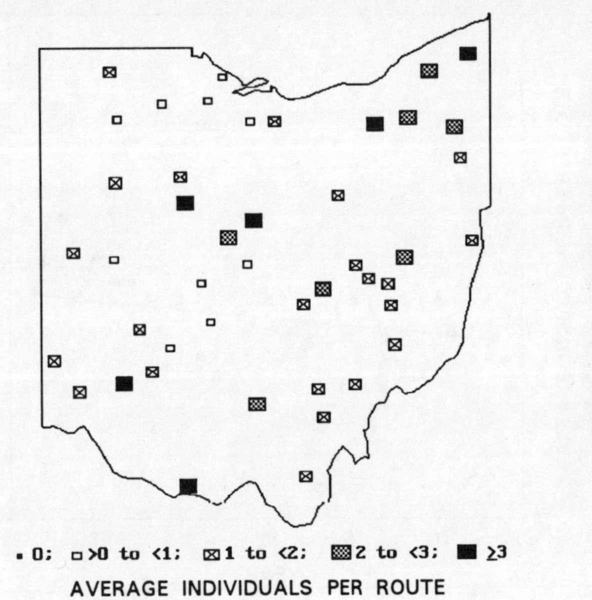

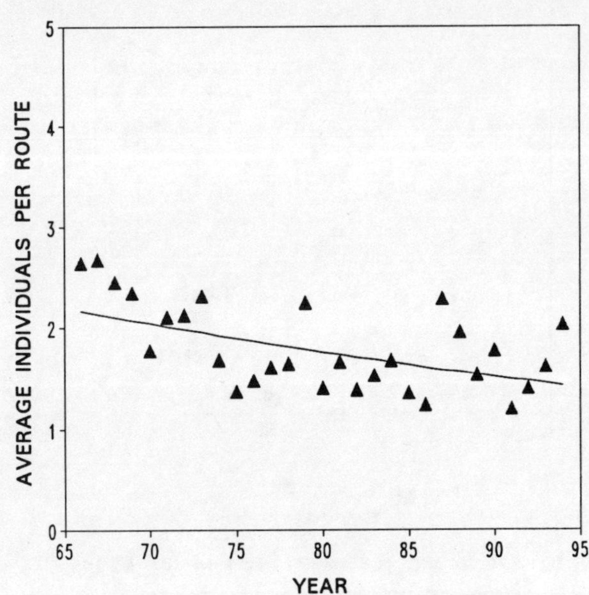

MIGRATORY STATUS: Central neotropical migrant.

BREEDING HABITAT: Mature, extensive woodlands probably preferred, but will also use smaller woodlots, young woods, and wooded riparian corridors. Nests in cavities.

ABUNDANCE AND DISTRIBUTION: Uncommon (1.8 birds per route) but widely distributed (all routes). Equally uncommon in Western and Eastern Ohio (1.6 vs. 2.0 birds per route, $P = 0.16$).

OHIO POPULATION TREND: Percent Annual Change ($\pm$ SE) = -1.5 ($\pm$ 0.6), P = 0.03
   Despite large variation around the regression line, Great Crested Flycatchers declined significantly at 1.5% annually. Trends in Western and Eastern Ohio were similar (-1.6 vs. -1.2% annually, $P = 0.71$).

CONTINENTAL AND GREAT LAKES REGIONAL TREND: Neither the regional nor continental population exhibited a significant 1966-94 trend (both -0.2%).

EASTERN KINGBIRD

Tyrannus tyrannus

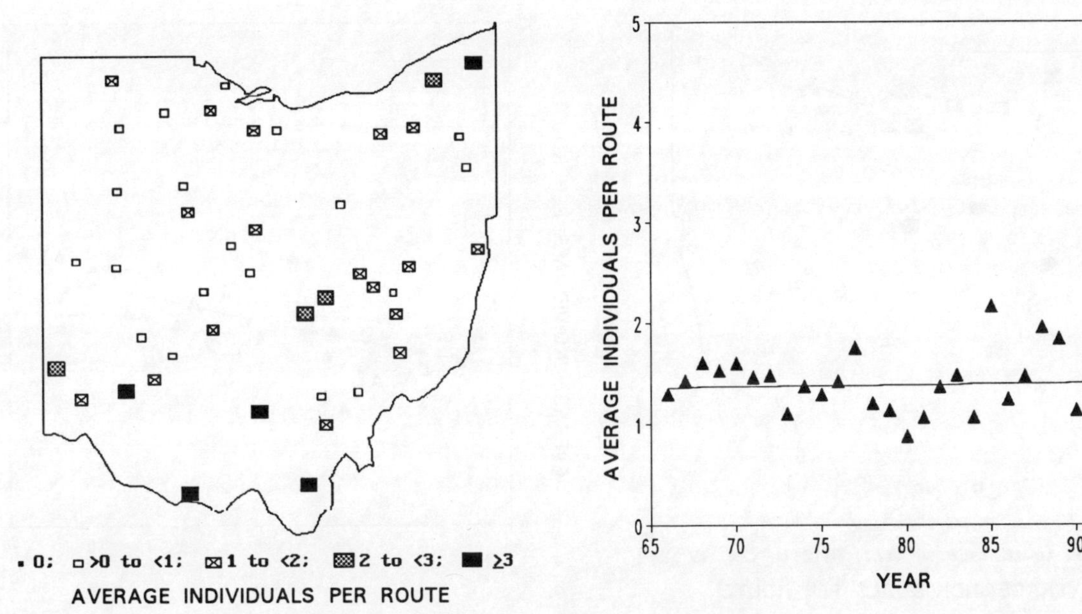

MIGRATORY STATUS: Southern neotropical migrant.

BREEDING HABITAT: Fairly open habitats where tall trees are interspersed with open areas.

ABUNDANCE AND DISTRIBUTION: Uncommon (1.4 birds per route) but widely distributed (all routes). Equally uncommon in Western and Eastern Ohio (1.2 vs. 1.5 birds per route, P = 0.32).

OHIO POPULATION TREND: Percent Annual Change (± SE) = 0.1 (± 0.8), P = 0.89
   Ohio's Eastern Kingbird population has been fairly stable since 1966 (0.1% annual change). Trends in Western and Eastern Ohio were not detectably different (-0.8 vs. 1.0%, P = 0.22).

CONTINENTAL AND GREAT LAKES REGIONAL TREND: Both the regional and continental populations have declined significantly since 1966 (1.6 and 0.4% annually).

HORNED LARK

Eremophila alpestris

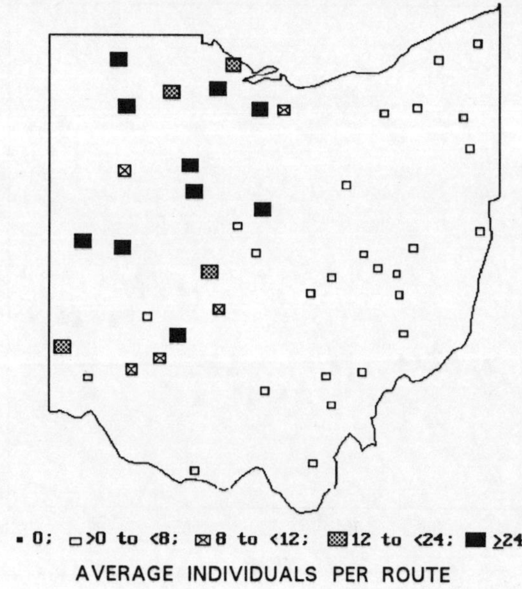

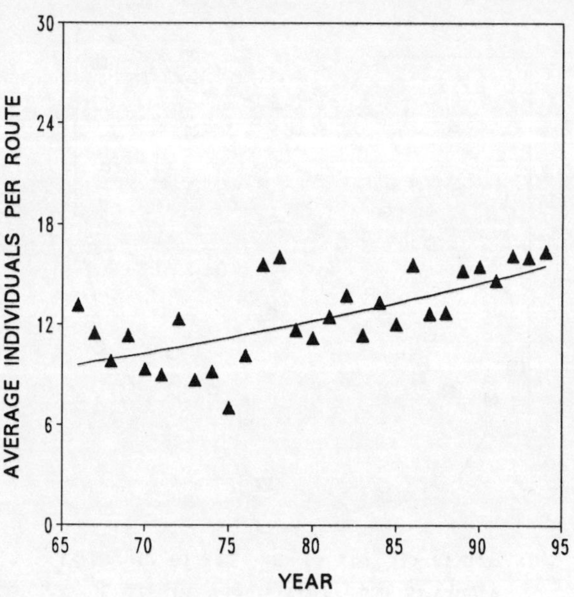

MIGRATORY STATUS: Permanent resident or temperate migrant. There is substantial winter movement into Ohio. Most individuals that breed in Ohio are thought to winter there also (Peterjohn, 1989).

BREEDING HABITAT: Grasslands and agricultural fields.

ABUNDANCE AND DISTRIBUTION: Common (12.6 birds per route) and widely distributed (all routes). Horned Larks were >11 times more common in predominantly agricultural Western Ohio than in heavily forested Eastern Ohio (21.9 vs. 1.9 birds per route, P < 0.001).

OHIO POPULATION TREND: **Percent Annual Change (± SE) = 1.7 (± 0.9), P = 0.06**
Horned Larks have increased modestly, and nearly significantly, at 1.7% annually. The increase appears to have occurred in Western Ohio (2.0%, P = 0.04), although the Eastern trend was highly variable (-2.2%, P = 0.36) and the trends were not statistically different from one another (P = 0.10).
Horned Larks colonized Ohio during the 1880s and were widespread by 1900. Apparently they have adapted well to changing land-use patterns and frequently forage in agriculture fields in Ohio.

CONTINENTAL AND GREAT LAKES REGIONAL TREND: The Great Lakes and continental populations have exhibited slight declines (-0.4 and -0.9% annually), but only the continental trend was significant.

PURPLE MARTIN

Progne subis

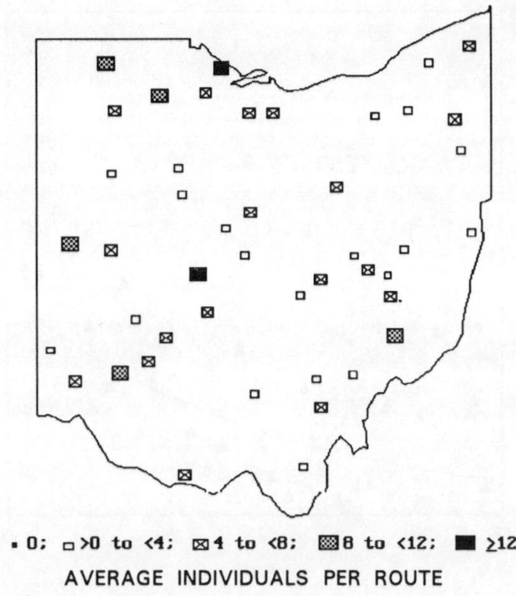

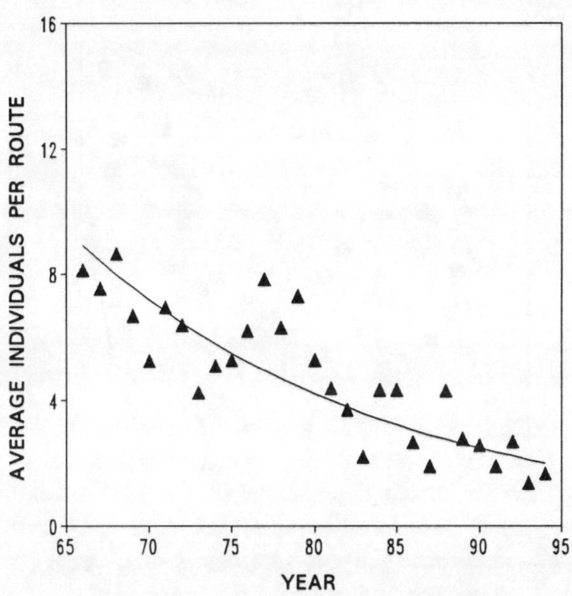

MIGRATORY STATUS: Southern neotropical migrant.

BREEDING HABITAT: In recent years, only known to nest in houses provided by humans. Preferred houses are located in residential areas near open fields and water.

ABUNDANCE AND DISTRIBUTION: Common (4.8 birds per route) and widely distributed (all routes). More common in Western than Eastern Ohio (5.9 vs. 3.4 birds per route P = 0.02).

OHIO POPULATION TREND: **Percent Annual Change (± SE) = -5.2 (± 1.1), P < 0.001**
   Ohio's Purple Martin population has declined sharply and significantly at 5.2% annually. Declines in Western and Eastern Ohio did not differ significantly (-6.2 vs. -3.6%, P = 0.26).
   Apparent declines during the late 1800s were blamed on competition with House Sparrows, but more recent declines are not thought to result from a shortage of nest sites (Peterjohn and Rice, 1991). The large annual fluctuations in Purple Martin populations are thought to result from high adult mortality and reproductive failure during cold, wet springs (Peterjohn, 1989).

CONTINENTAL AND GREAT LAKES REGIONAL TREND: Like the Ohio population, the regional population decreased significantly at 3.8% annually. The continental population did not exhibit a significant change (0.1%).

TREE SWALLOW

Tachycineta bicolor

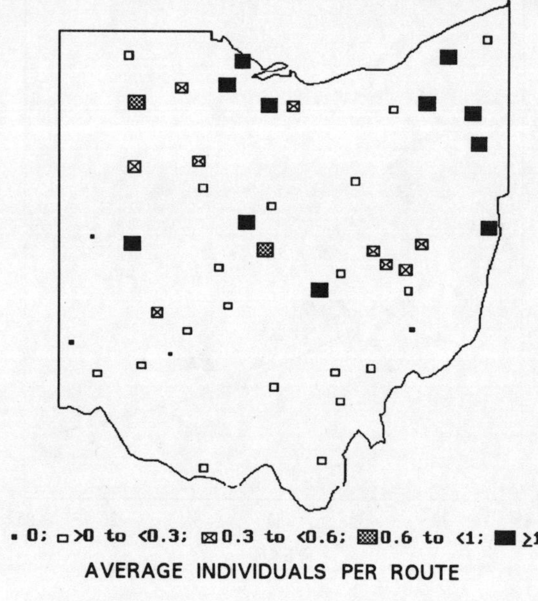

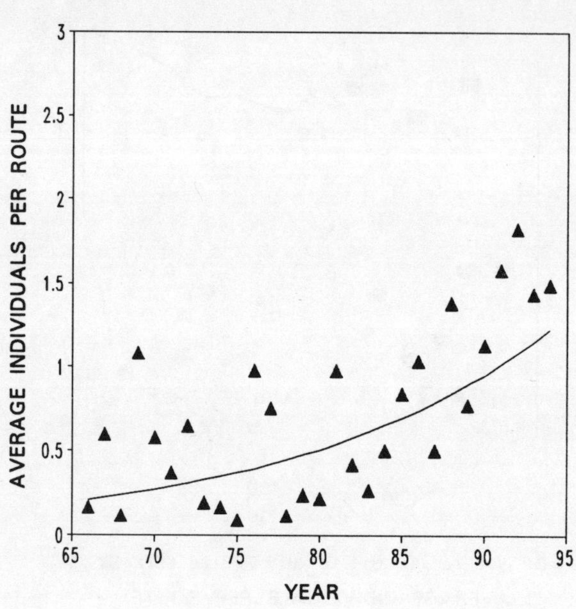

MIGRATORY STATUS: Temperate migrant.

BREEDING HABITAT: Wooded wetlands or wetlands where nest boxes have been erected.

ABUNDANCE AND DISTRIBUTION: Rarely recorded (0.70 birds per route) but widely distributed (41 routes). Ohio is on the southern edge of the breeding range, and Tree Swallows were more common in Northern than Southern Ohio (1.2 vs. 0.3 birds per route, $P = 0.007$).

OHIO POPULATION TREND: Percent Annual Change ($\pm$ SE) = 6.5 ($\pm$ 1.8), P = 0.001
Ohio's Tree Swallow population increased significantly at 6.5% annually, despite large year-to-year variation in numbers recorded. Substantial year-to-year variation is due to the presence of large flocks on a few routes in some years and not others.
The increase was more pronounced in Southern than Northern Ohio (12.0 vs. 6.1%, $P = 0.05$), suggesting range expansion into Southern Ohio.

CONTINENTAL AND GREAT LAKES REGIONAL TREND: Like the Ohio population, the regional population increased significantly at 2.0% annually; the continental population did not exhibit a significant annual change (0.6%).

NORTHERN ROUGH-WINGED SWALLOW

Stelgidopteryx serripennis

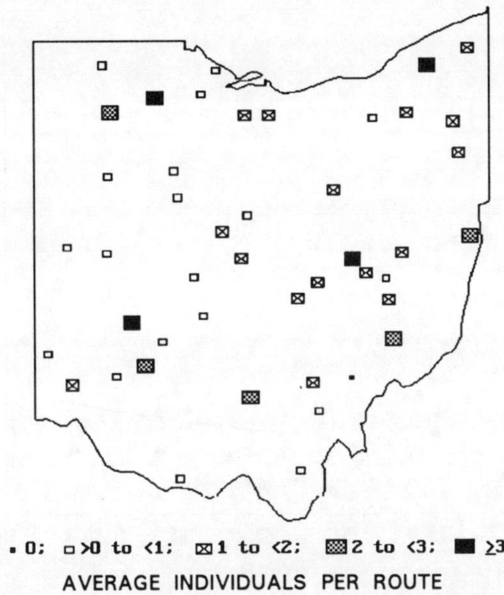

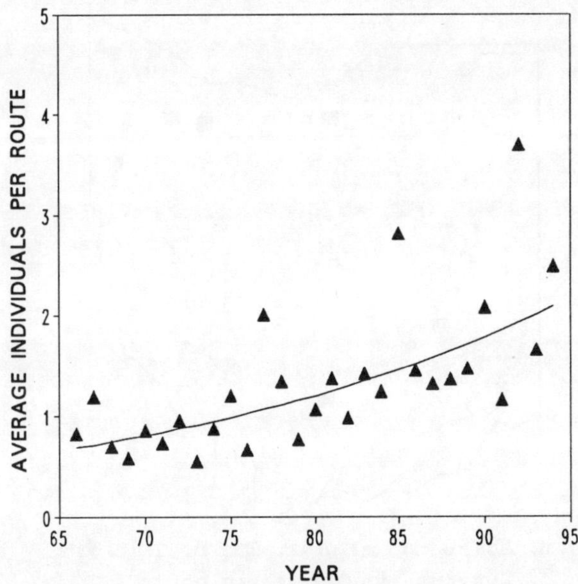

MIGRATORY STATUS: Southern neotropical migrant.

BREEDING HABITAT: Riparian corridors. Nest solitarily or in small colonies.

ABUNDANCE AND DISTRIBUTION: Uncommon (1.3 birds per route) and widely distributed (44 routes). Abundance in Western and Eastern Ohio not detectably different (1.2 vs. 1.5 birds per route, P = 0.29).

OHIO POPULATION TREND: Percent Annual Change ($\pm$ SE) = 4.0 ($\pm$ 0.9), P < 0.001
   Northern Rough-winged Swallows increased significantly in Ohio at 4.0% annually. The apparent outliers in 1977, 1985, and 1992 are due to flocks of >10 birds on 1-2 routes in each year. If the flocks are excluded, the trend changes little (3.9%, P < 0.001).
   Trends in Eastern and Western Ohio were similar (3.6 vs. 5.1%, P = 0.38).

CONTINENTAL AND GREAT LAKES REGIONAL TREND: Neither the regional nor continental populations exhibited significant annual changes (0.2 and -0.3%).

BANK SWALLOW

Riparia riparia

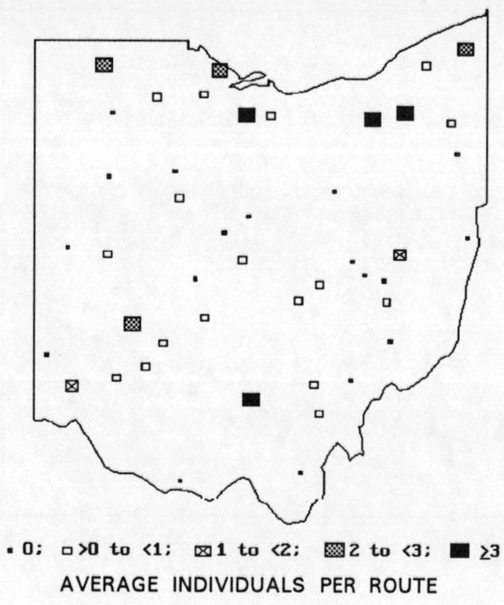

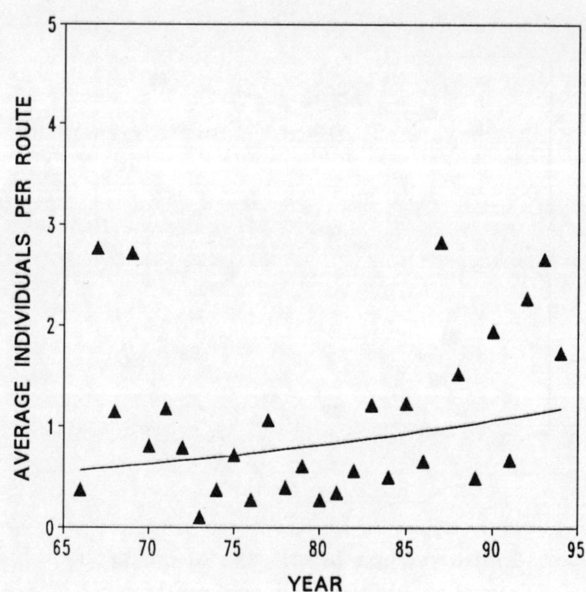

MIGRATORY STATUS: Southern neotropical migrant.

BREEDING HABITAT: Large rivers and lakes where steep banks are available for nesting. Also nest in quarries. Typically nest in large colonies.

ABUNDANCE AND DISTRIBUTION: Rare (1.0 birds per route) and locally distributed (27 routes). Abundance in Western and Eastern Ohio not detectably different (0.9 vs. 1.2 birds per route, $P = 0.69$). Somewhat more abundant in Northern than Southern Ohio (1.8 vs. 0.4 birds per route, $P = 0.09$).

OHIO POPULATION TREND: Percent Annual Change ($\pm$ SE) = 2.7 ($\pm$ 3.2), $P = 0.41$

Bank Swallows did not exhibit a significant overall trend (2.7% annual change). The large annual variation in number recorded per route would make a long-term trend difficult to detect. The large annual variation is probably due to the chance encounter of large flocks that forage across extensive areas and thus are often absent when a given route is run.

Trends in Western vs. Eastern Ohio (-0.9 vs. 5.9%, $P = 0.30$) and Northern vs. Southern Ohio (2.4 vs. 3.7%, $P = 0.81$) did not differ significantly.

CONTINENTAL AND GREAT LAKES REGIONAL TREND: Similarly, neither the regional nor continental populations exhibited significant annual changes (-1.1 and -1.1%).

BARN SWALLOW

Hirundo rustica

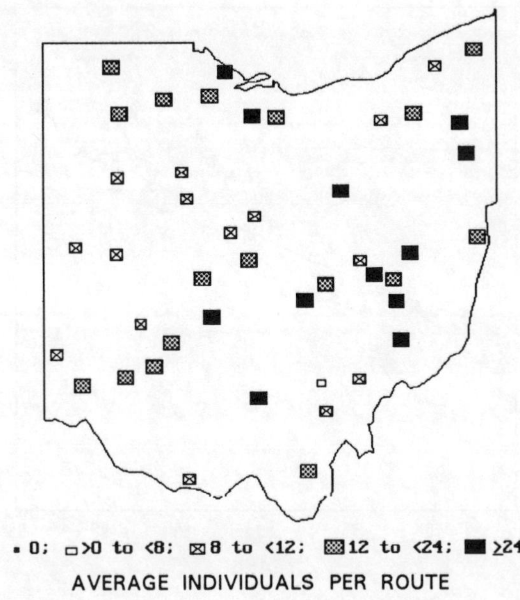

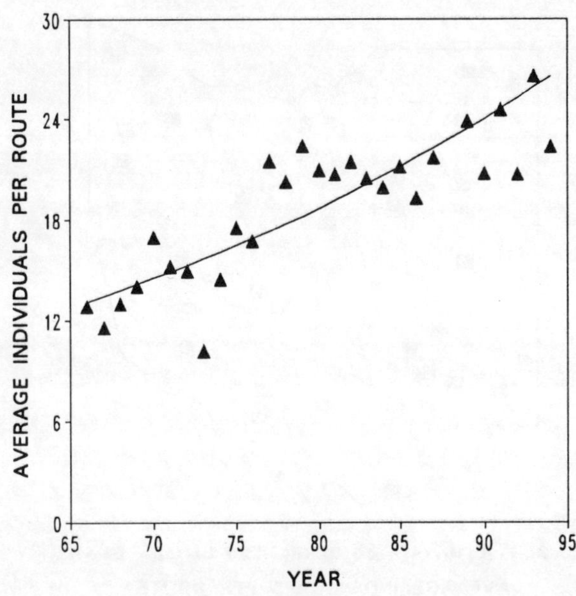

MIGRATORY STATUS: Southern neotropical migrant.

BREEDING HABITAT: Residential areas, particularly rural areas where they nest in barns and forage over open fields. Nest solitarily or in small colonies.

ABUNDANCE AND DISTRIBUTION: Abundant (19.4 birds per route) and widely distributed (all routes). Slightly more common in Eastern than Western Ohio (22.1 vs. 17.0 birds per route, P = 0.06).

OHIO POPULATION TREND: Percent Annual Change (± SE) = 2.5 (± 0.5), P < 0.001
   Barn Swallows have increased significantly at a marked 2.5% annual rate. Trends in Eastern and Western Ohio did not differ significantly (1.9 vs. 3.3%, P = 0.19).

CONTINENTAL AND GREAT LAKES REGIONAL TREND: The regional population increased slightly, but significantly, at 0.5% annually, and the continental population exhibited no significant trend (-0.1%).

BLUE JAY

Cyanocitta cristata

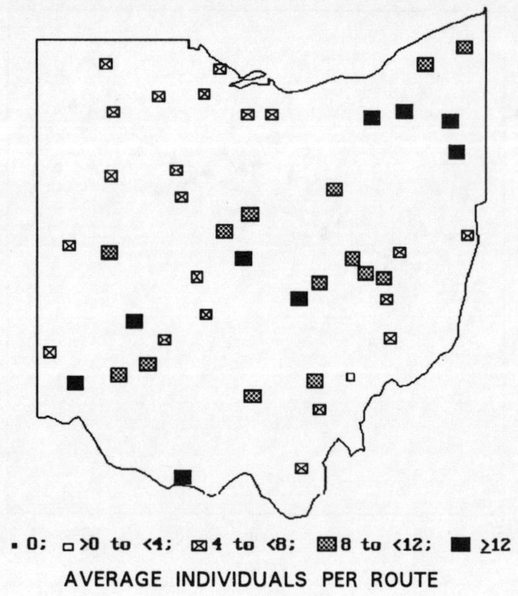

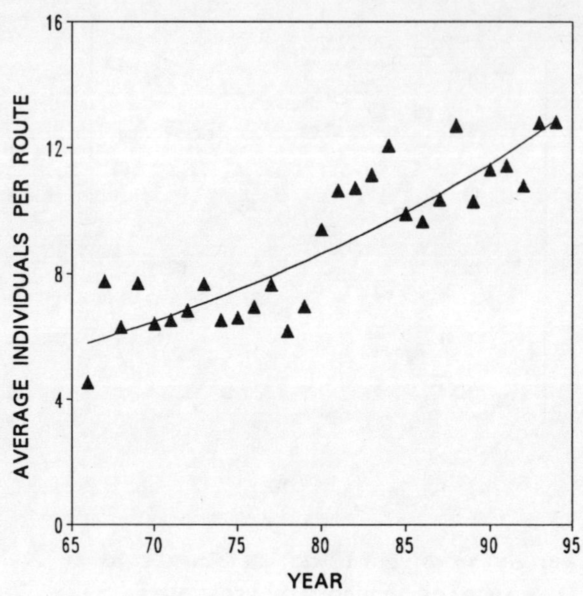

MIGRATORY STATUS: Permanent resident or temperate migrant. There is substantial migratory movement, but many individuals winter in Ohio.

BREEDING HABITAT: Woodlands, especially young woods, edges, and corridors.

ABUNDANCE AND DISTRIBUTION: Common (9.0 birds per route) and widely distributed (all routes). Abundance in Eastern and Western Ohio not significantly different (10.0 vs. 8.2 birds per route, $P = 0.17$).

OHIO POPULATION TREND: Percent Annual Change ($\pm$ SE) = 2.9 ($\pm$ 0.5), P < 0.001
  Blue Jays have increased significantly at a striking 2.9% annual rate. Trends in Eastern and Western Ohio were similar (3.2 vs. 2.7%, $P = 0.66$).

CONTINENTAL AND GREAT LAKES REGIONAL TREND: In contrast to Ohio's population, the Great Lakes regional and continental populations have decreased significantly (0.5 and 1.6% annually).

AMERICAN CROW

Corvus brachyrhynchos

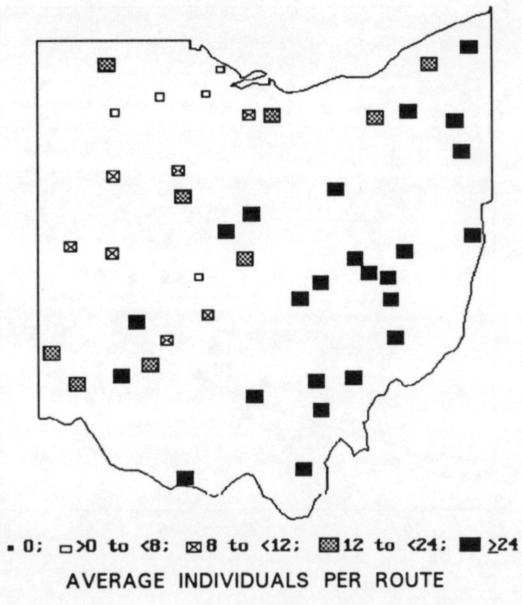

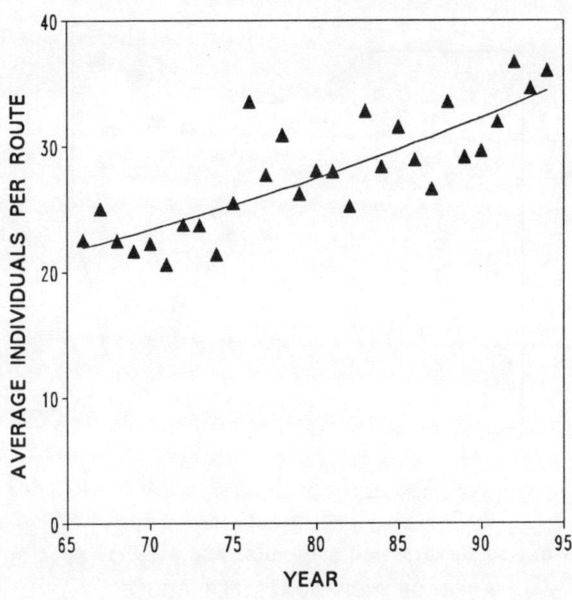

MIGRATORY STATUS: Permanent resident or temperate migrant. Many individuals winter in Ohio; it is not known whether individuals that breed in Ohio also winter there.

BREEDING HABITAT: Woodlands, especially those interspersed with agricultural and early successional fields.

ABUNDANCE AND DISTRIBUTION: Abundant (27.8 birds per route) and widely distributed (all routes). More than twice as common in Eastern than Western Ohio (38.8 vs. 18.1 birds per route, P < 0.001).

OHIO POPULATION TREND: **Percent Annual Change (± SE) = 1.6 (± 0.4), P = 0.001**

American Crows increased significantly in Ohio at 1.6% annually. Trends in Eastern and Western Ohio were similar (1.9 vs. 1.3%, P = 0.46).

Historically, Ohio's population expanded as extensive virgin forests were fragmented into a mosaic of farmlands and woodlots. However, since the early 1950s, the landscape in western counties has become less favorable to crows because intensive farming has decreased the availability of woodlots (Peterjohn, 1989).

CONTINENTAL AND GREAT LAKES REGIONAL TREND: Similarly, the regional and continental populations have also increased modestly, but significantly, at 1.0 and 0.7% annually.

BLACK-CAPPED CHICKADEE

Parus atricapillus

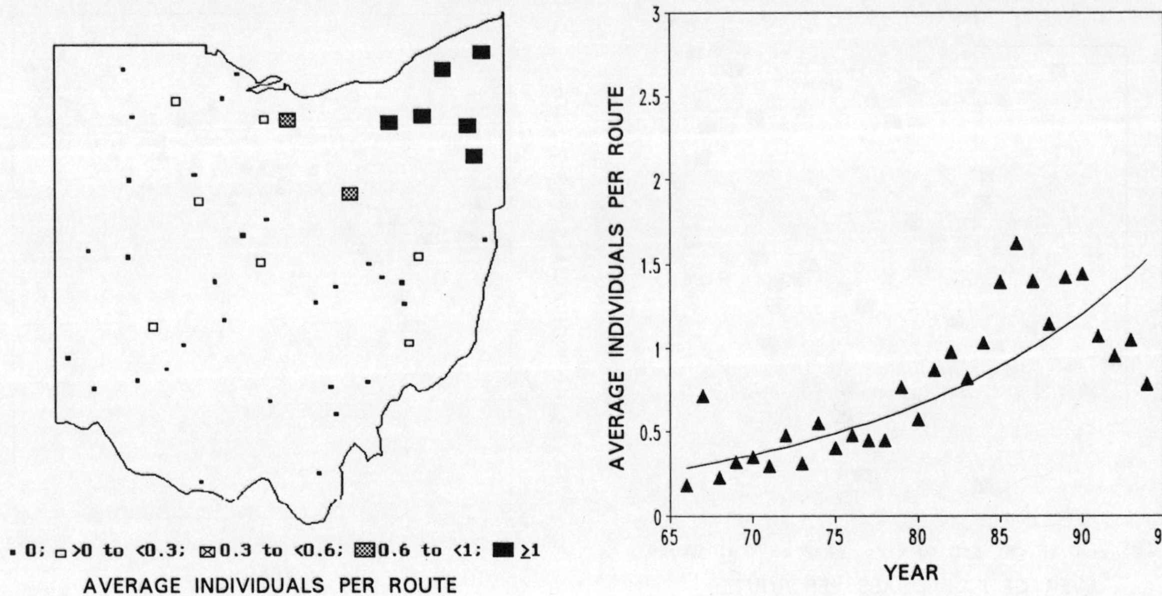

MIGRATORY STATUS: Permanent resident.

BREEDING HABITAT: Woodlands, including mature and young woodlands and edges.

ABUNDANCE AND DISTRIBUTION: Typical breeding range includes only northern Ohio. Within northern Ohio, Black-capped Chickadees are uncommonly recorded (1.5 birds per route). Black-capped Chickadees are probably underrepresented, relative to other species, in BBS data because they rarely vocalize in early June.

OHIO POPULATION TREND: Percent Annual Change (± SE) = 6.2 (± 1.3), P < 0.001
  Black-capped Chickadees increased significantly and steadily at a dramatic annual rate of 6.2%. The population appears to have declined since 1985 at 6.1% (P = 0.33), although the trend is not statistically significant because most chickadees are observed on only 6 of the 45 routes. Occurred on too few routes to allow comparisons of regional trends.

CONTINENTAL AND GREAT LAKES REGIONAL TREND: Similarly, the regional and continental populations have also increased significantly at 2.1 and 1.5% annually.

CAROLINA CHICKADEE

Parus carolinensis

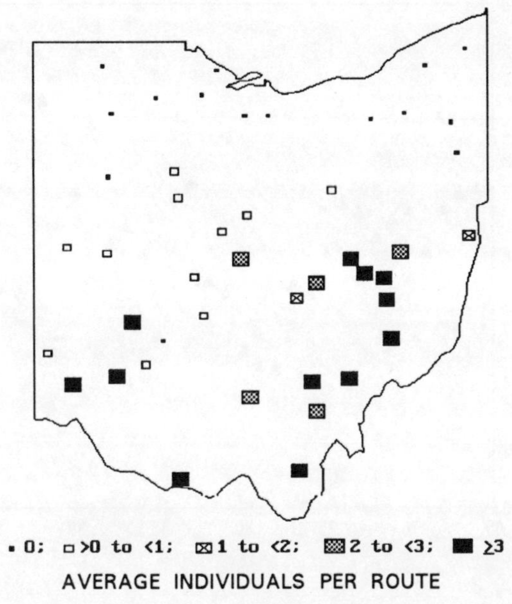

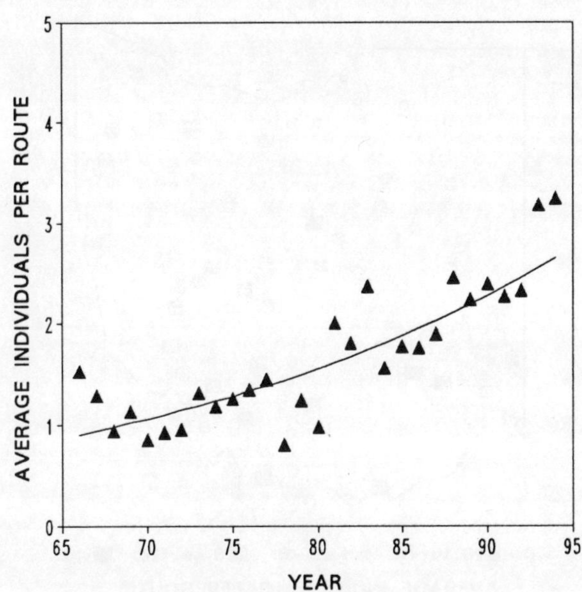

MIGRATORY STATUS: Permanent resident.

BREEDING HABITAT: Woodlands, including mature and young woodlands and edges.

ABUNDANCE AND DISTRIBUTION: Typical breeding range includes only southern Ohio. Within southern Ohio, Carolina Chickadees are fairly common (2.9 birds per route). Like Black-capped Chickadees, Carolina Chickadees are probably relatively more common than indicated by BBS because their peak of vocalization occurs in early spring and they vocalize little in early June.

OHIO POPULATION TREND: Percent Annual Change ($\pm$ SE) = 3.9 ($\pm$ 0.8), P < 0.001
   Carolina Chickadees increased significantly at 3.9% annually. Trends in Eastern and Western Ohio did not differ significantly (3.2 vs. 5.0%, P = 0.35).

CONTINENTAL AND GREAT LAKES REGIONAL TREND: In contrast, the continental population declined significantly at 0.8% annually and the Great Lakes population exhibited no signficant overall trend (-0.1%).

TUFTED TITMOUSE

Parus bicolor

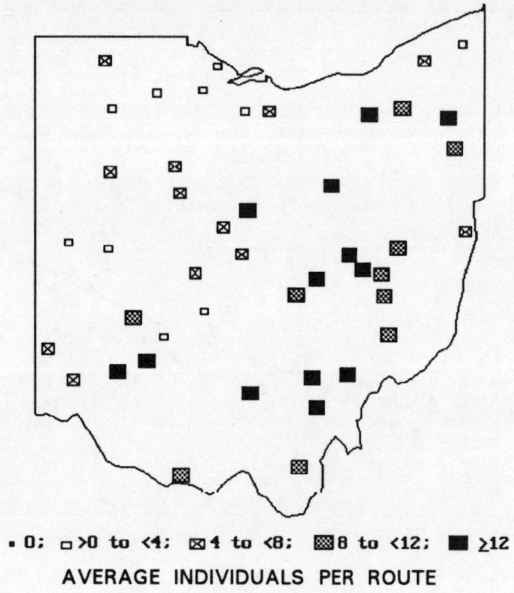

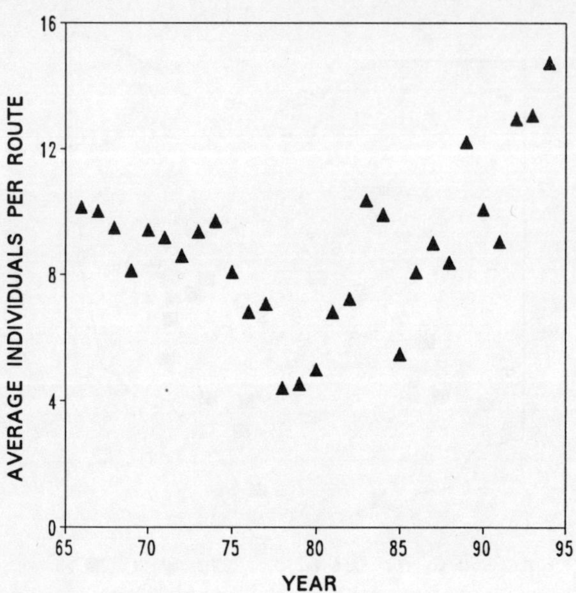

MIGRATORY STATUS: Permanent resident.

BREEDING HABITAT: Woodlands, including mature forests and others with scattered large trees suitable for nest excavation.

ABUNDANCE AND DISTRIBUTION: Common (8.9 birds per route) and widely distributed (all routes). Like other parids, Tufted Titmice vocalize little in early June and are probably underrepresented in BBS data relative to other species.

Nearly twice as common in Eastern than Western Ohio (11.9 vs. 6.2 birds per route, P < 0.001).

OHIO POPULATION TREND: Percent Annual Change calculated for 1966-1994 not appropriate.

Tufted Titmice declined sharply during 1966-1978 at 4.3% annually (P < 0.001), due in part to the severe winters of 1976-1978, and then increased significantly at 5.6% from 1979-1994 (P < 0.001). Similar to Carolina Chickadees, their numbers exhibit substantial annual variation which is probably due to variation in winter weather and subsequent adult mortality. Trends in Eastern and Western Ohio were similar (0.9 vs. 0.7% annual change, P = 0.88).

CONTINENTAL AND GREAT LAKES REGIONAL TREND: The regional population exhibited no significant overall trend (0.5%) and the continental population increased significantly at 0.8% annually. Note that the Tufted Titmouse regional and continental trends may be misleading if the extreme winter effect observed in Ohio was evident elsewhere.

WHITE-BREASTED NUTHATCH

Sitta carolinensis

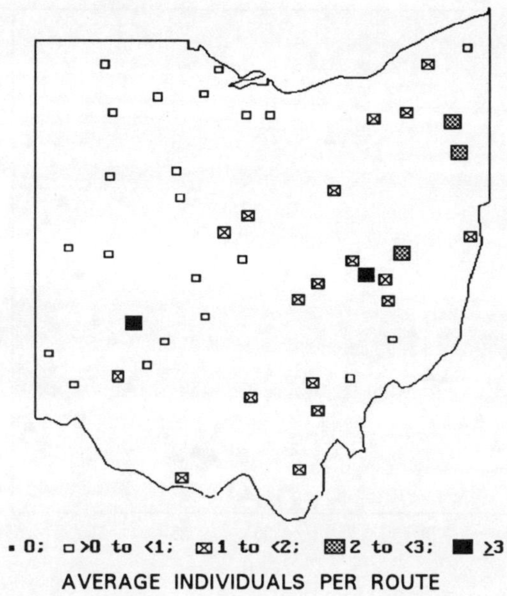

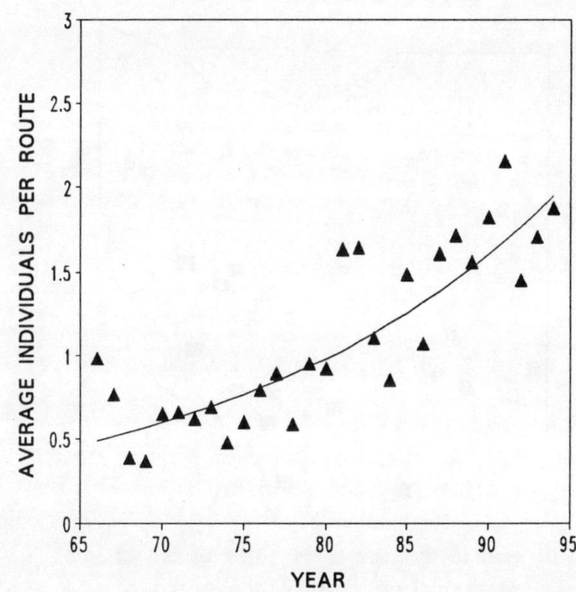

MIGRATORY STATUS: Permanent resident.

BREEDING HABITAT: Woodlands, including mature forests and others with scattered large trees.

ABUNDANCE AND DISTRIBUTION: Rarely recorded (1.1 birds per route) and widely distributed (all routes). Because White-breasted Nuthatches vocalize less than most bird species and are difficult to detect visually, this species is probably more common, relative to other species, than indicated by BBS data.

More than twice as common in Eastern than Western Ohio (1.6 vs. 0.7 birds per route, P < 0.001).

OHIO POPULATION TREND: Percent Annual Change ($\pm$ SE) = 5.0 ($\pm$ 0.6), P < 0.001

White-breasted Nuthatches increased significantly at 5.0% annually. Although increasing significantly in both Eastern (4.1%, P < 0.001) and Western Ohio (8.5%, P < 0.001), the rate of increase in Western Ohio was significantly greater than in Eastern Ohio (P = 0.006).

CONTINENTAL AND GREAT LAKES REGIONAL TREND: Similarly, the regional and continental populations also increased significantly at 1.5 and 2.1% annually.

CAROLINA WREN

Thryothorus ludovicianus

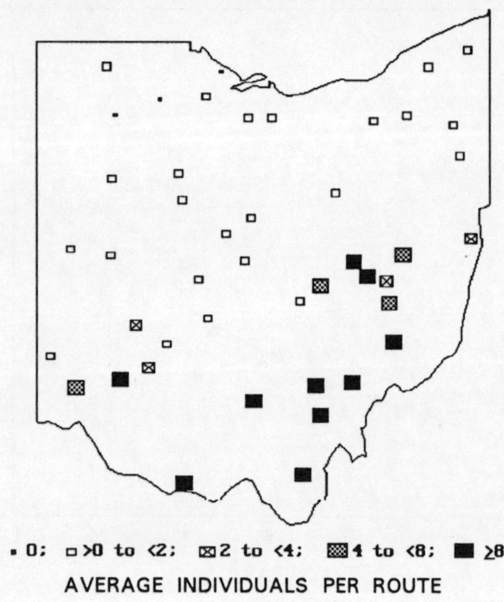

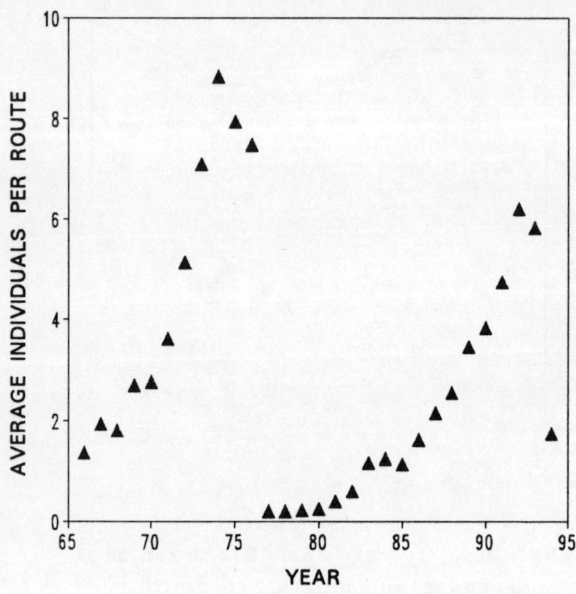

MIGRATORY STATUS: Permanent resident.

BREEDING HABITAT: Scrub, brushy edges, and young woods with dense undergrowth.

ABUNDANCE AND DISTRIBUTION: Fairly common (3.0 birds per route) and widely distributed (42 routes). Typically more common in the southeastern U.S. than elsewhere, Carolina Wrens were also more common in Southern than Northern Ohio (5.2 vs. 0.6 birds per route, P < 0.001).

OHIO POPULATION TREND: Percent Annual Change calculated for 1966-1994 not appropriate.
Ohio's Carolina Wren population increased dramatically at 22% annually during 1966-1976 (P < 0.001), was decimated by the severe winters of 1976-77 and 1977-78, and has again increased dramatically at 23% during 1979-1994 (P < 0.001).

CONTINENTAL AND GREAT LAKES REGIONAL TREND: The regional population exhibited no significant trend (1.9%) and the continental population increased significantly at 0.9% annually. Note that the Carolina Wren's regional and continental trends may be misleading if the extreme winter effect observed in Ohio was evident elsewhere.

HOUSE WREN

Troglodytes aedon

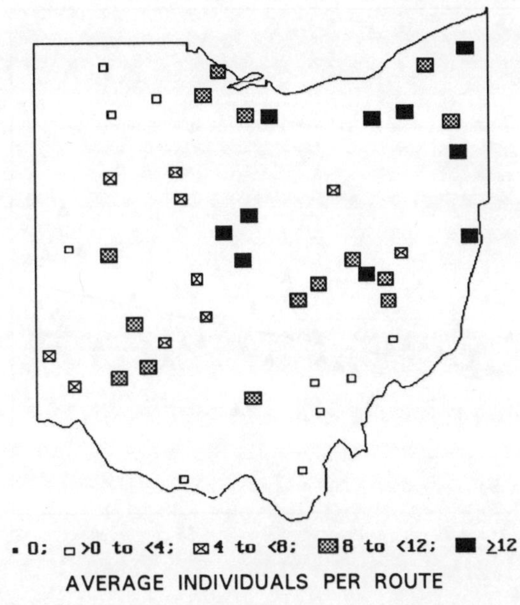

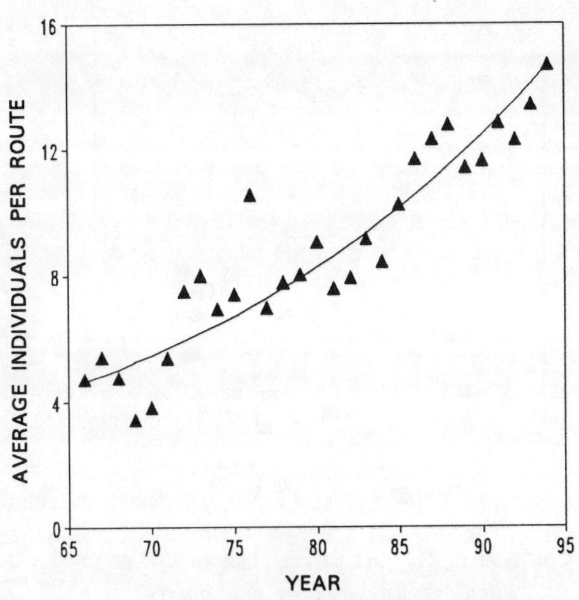

MIGRATORY STATUS: Temperate migrant.

BREEDING HABITAT: Woodlands, especially edges and young stands or openings.

ABUNDANCE AND DISTRIBUTION: Common (8.8 birds per route) and widely distributed (all routes). Abundance in Western and Eastern Ohio not statistically different (8.1 vs. 9.7 birds per route, P = 0.31).

OHIO POPULATION TREND: Percent Annual Change ($\pm$ SE) = 4.2 ($\pm$ 0.7), P < 0.001

House Wrens have increased significantly in Ohio at 4.2% annually. Increases in Western and Eastern Ohio were similar (4.6 vs. 3.8%, P = 0.54).

Historically, the distribution of House Wrens probably expanded as Ohio was settled. Populations may have declined during the late 1800s but had recovered by the 1930s (Peterjohn, 1989).

CONTINENTAL AND GREAT LAKES REGIONAL TREND: Similarly, the regional and continental populations have increased significantly at 1.3 and 1.6% annually.

BLUE-GRAY GNATCATCHER

Polioptila caerulea

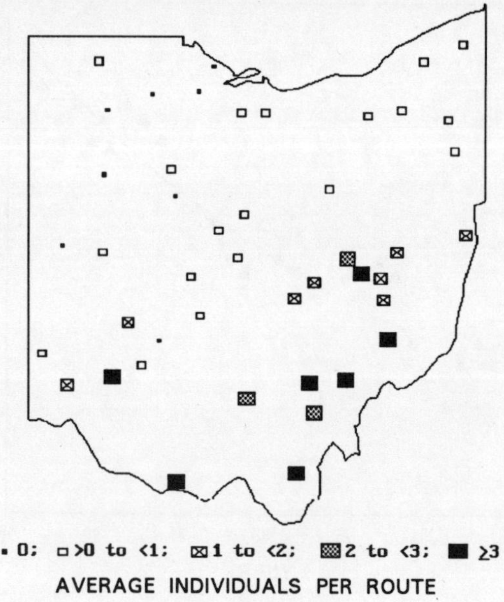

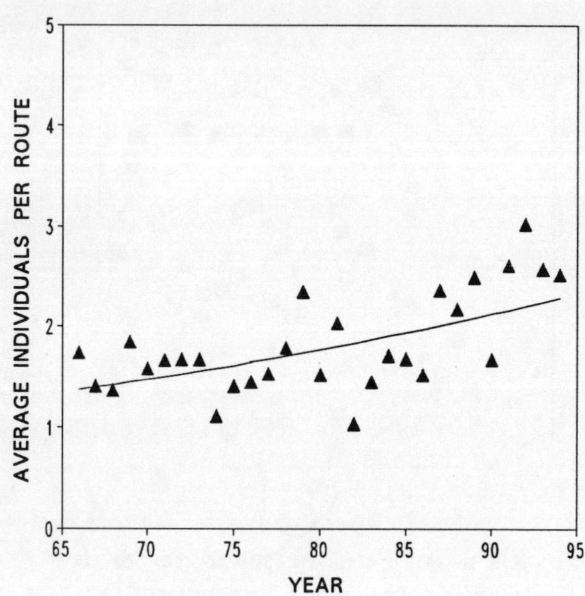

MIGRATORY STATUS: Central neotropical migrant.

BREEDING HABITAT: Mature woods.

ABUNDANCE AND DISTRIBUTION: Uncommon (1.8 birds per route) and fairly widely distributed (37 routes). Abundance in Eastern and Western Ohio did not differ significantly (2.6 vs. 1.1 birds per route, P = 0.09). Typically more common in the southern U.S., Blue-gray Gnatcatchers were also more abundant in Southern than Northern Ohio (2.8 vs. 0.4 birds per route, P = 0.002).

OHIO POPULATION TREND: Percent Annual Change ($\pm$ SE) = 1.9 ($\pm$ 0.9), P = 0.05

   Blue-gray Gnatcatchers exhibited a nearly significant population increase of 1.9% annually. Blue-gray Gnatcatchers have expanded their range northward in eastern North America since 1965 (Robbins et al., 1986). This expansion was evident in Northern Ohio beginning in 1970 (Peterjohn, 1989). However, the Northern and Southern Ohio population trends for 1966-1994 did not differ significantly (2.3 vs. 1.6%, P = 0.80), in part because of high annual variability in both regions.

CONTINENTAL AND GREAT LAKES REGIONAL TREND: The regional trend was not significant (0.5%), but the continental population exhibited a significant increase at 1.2% annually.

EASTERN BLUEBIRD

Sialia sialis

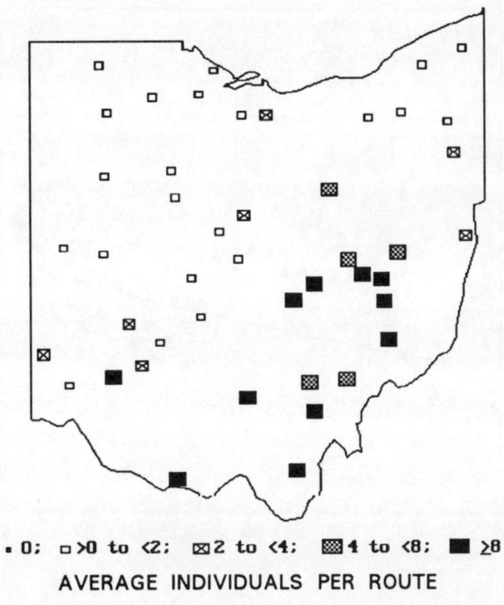

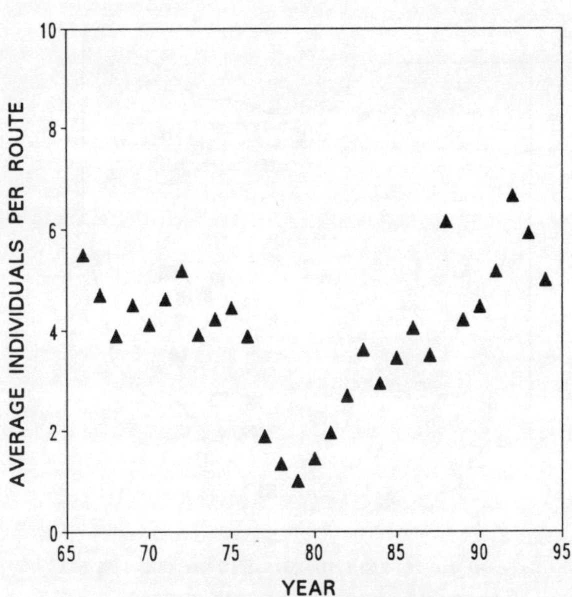

MIGRATORY STATUS: Temperate migrant.

BREEDING HABITAT: Grasslands wherever boxes or natural cavities are available for nests.

ABUNDANCE AND DISTRIBUTION: Common (4.0 birds per route) and widely distributed (all routes). Much more common in Eastern than Western Ohio (6.3 vs. 1.9 birds per route, P < 0.001). In the intensively farmed western counties, most natural cavities have been eliminated and bluebirds are largely dependent on nest boxes (Peterjohn, 1989).

OHIO POPULATION TREND: Percent Annual Change calculated for 1966-1994 not appropriate.
  Numbers of Eastern Bluebirds in Ohio were fairly stable at -1.6% annual change from 1966-1976 (P = 0.15), declined sharply after the severe winters of 1976-1978, and recovered steadily at 10.3% thereafter (1979-1994, P < 0.001). Eastern Bluebirds increased somewhat, but not significantly, in Western Ohio (2.5%, P = 0.08), where they are less common, but not in Eastern Ohio (-0.1%, P = 0.89).
  Historically, the population probably increased as Ohio was settled and more open grassy fields were available. Population declines and recoveries, similar to those after the 1976-1978 winters, occurred after the spring of 1895 and after the winters of 1958-60 (Peterjohn, 1989).

CONTINENTAL AND GREAT LAKES REGIONAL TREND: The regional and continental populations increased significantly at 1.9 and 2.3% annually. Note that the Eastern Bluebird's regional and continental trends may be misleading if the extreme winter effect observed in Ohio was evident elsewhere.

WOOD THRUSH

Hylocichla mustelina

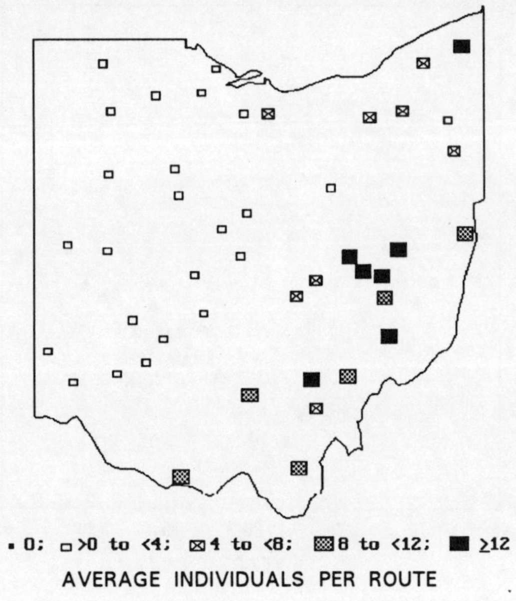

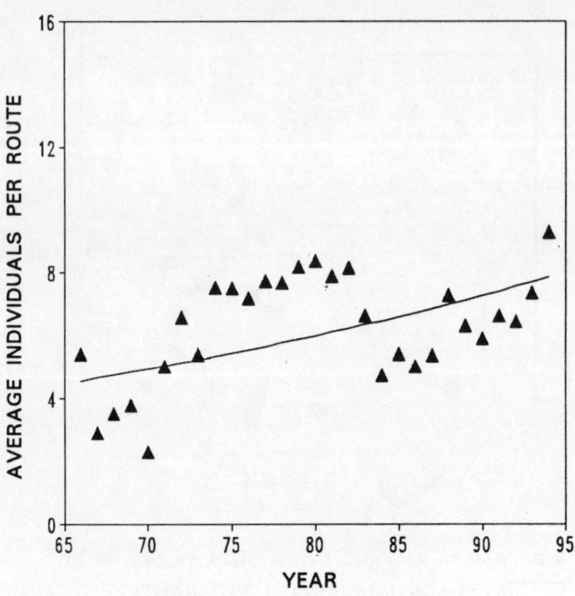

MIGRATORY STATUS: Central neotropical migrant.

BREEDING HABITAT: Woodlands, including young and mature woods, especially those with a moderately dense understory.

ABUNDANCE AND DISTRIBUTION: Common (6.3 birds per route) and widely distributed (all routes). Wood Thrush abundance is >6 times greater in Eastern than Western Ohio (11.5 vs. 1.9 birds per route, P < 0.001).

OHIO POPULATION TREND: **Percent Annual Change (± SE) = 2.0 (± 0.5), P < 0.001**
   The Wood Thrush population is best described as having increased at 6.6% through 1982 (P = 0.002), decreased sharply in 1983 and 1984, and increased somewhat, but not significantly, through 1984-1994 at 5.2% annually (P = 0.17). The overall 1966-1994 trend is increasing and significant (2.0%). The increases in Eastern and Western Ohio did not differ significantly (3.3 vs. 1.7% annually, P = 0.22).

CONTINENTAL AND GREAT LAKES REGIONAL TREND: In contrast to Ohio's population, the continental population decreased significantly at 1.8% annually, and the Great Lakes population exhibited no significant trend (-0.6%).

AMERICAN ROBIN

Turdus migratorius

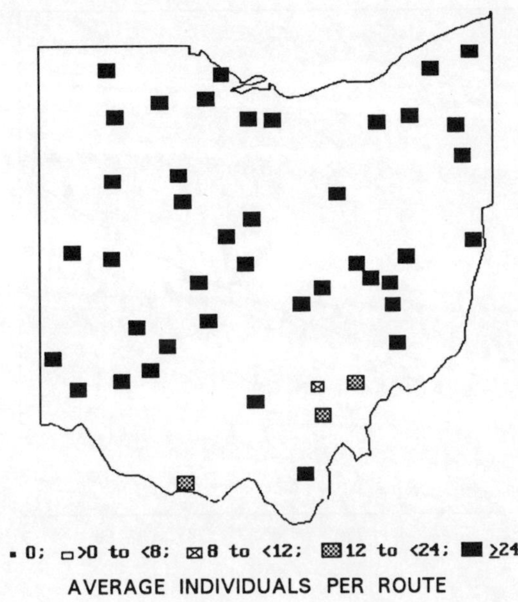

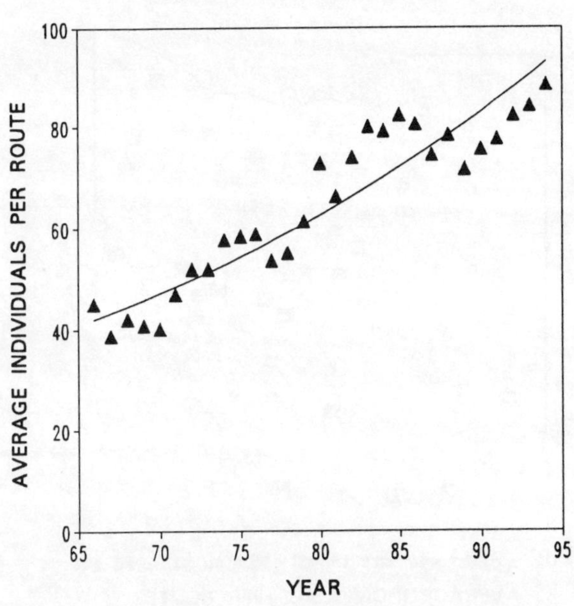

MIGRATORY STATUS: Permanent resident, although there is substantial winter movement.

BREEDING HABITAT: A generalist, often in suburban areas, woodland borders or openings.

ABUNDANCE AND DISTRIBUTION: Very abundant (64.4 birds per route) and widely distributed (all routes). Equally abundant in Western and Eastern Ohio (67.1 vs. 61.2 birds per route, $P = 0.44$).

OHIO POPULATION TREND: Percent Annual Change ($\pm$ SE) = 2.9 ($\pm$ 0.5), $P < 0.001$
  The increase in American Robins was highly significant and consistent throughout the period (2.9% annual change). The population increase was significantly greater in Western than Eastern Ohio (3.8 vs. 1.7%, $P = 0.04$).

CONTINENTAL AND GREAT LAKES REGIONAL TREND: Similarly, the regional and continental populations increased significantly at 2.0 and 0.8% annually. Robbins et al. (1986) suggested that the continent-wide increase may have been a response, in part, to a decline in pesticide contamination.

GRAY CATBIRD

Dumetella carolinensis

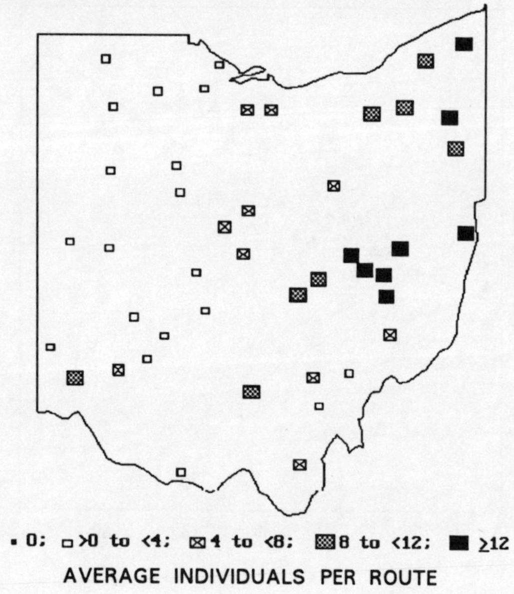

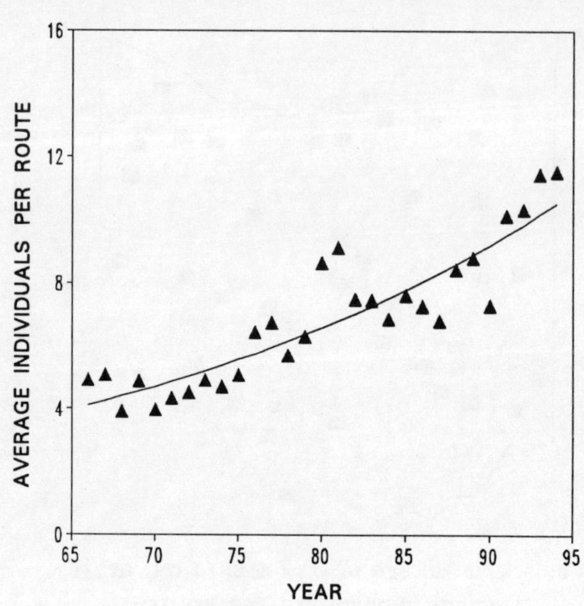

MIGRATORY STATUS: Central neotropical migrant.

BREEDING HABITAT: Scrub, including early successional habitat, edges, and disturbed woodlands.

ABUNDANCE AND DISTRIBUTION: Common (7.0 birds per route) and widely distributed (all routes). Nearly 3 times more common in Eastern than Western Ohio (10.8 vs. 3.6 birds per route, $P < 0.001$).

OHIO POPULATION TREND: Percent Annual Change ($\pm$ SE) = 3.4 ($\pm$ 0.8), $P < 0.001$
   Gray Catbirds have increased significantly, and fairly steadily, at 3.4% annually. Trends in Eastern and Western Ohio were similar (3.3 vs. 4.0%, $P = 0.63$).

CONTINENTAL AND GREAT LAKES REGIONAL TREND: In contrast to Ohio's population, the continental population decreased significantly at 0.4% annually, and the regional population exhibited no significant trend (0.4%).

NORTHERN MOCKINGBIRD

Mimus polyglottos

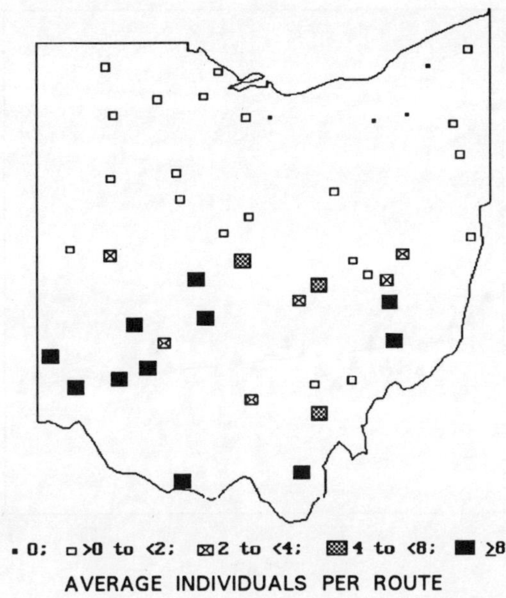

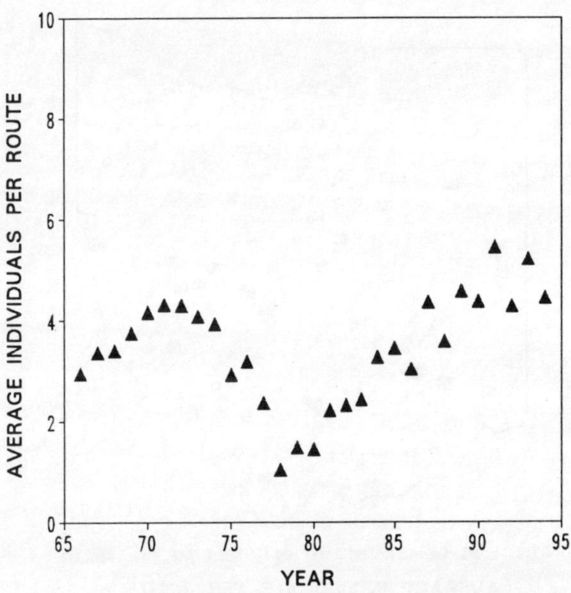

MIGRATORY STATUS: Resident, in general, but a few individuals move south during winter.

BREEDING HABITAT: Scrub, especially brushy thickets near human habitation.

ABUNDANCE AND DISTRIBUTION: Fairly common (3.4 birds per route) and widely distributed (41 routes). Much more common in Southern than Northern Ohio (6.0 vs. 0.4 birds per route, P < 0.001).

OHIO POPULATION TREND: Percent Annual Change calculated for **1966-1994 not appropriate.**
    Northern Mockingbird numbers declined sharply during the severe winters of 1976-1978 and have recovered steadily since then at a significant rate of 8.4% annually (P = 0.001). Trends in Southern and Northern Ohio were not significantly different (0.8 vs. 0.2%, P = 0.70).
    Mockingbirds began spreading northward into Southern Ohio in the mid-1800s and expanded to Northern Ohio in the 1930s-1940s (Peterjohn, 1989).

CONTINENTAL AND GREAT LAKES REGIONAL TREND: The regional and continental populations have declined significantly at 2.1 and 1.0% annually. Note that the Northern Mockingbird's regional and continental trends may be misleading if the extreme winter effect observed in Ohio was evident elsewhere.

BROWN THRASHER

Toxostoma rufum

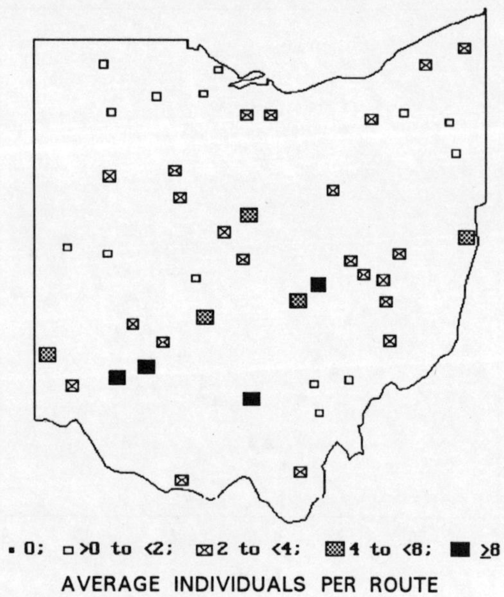

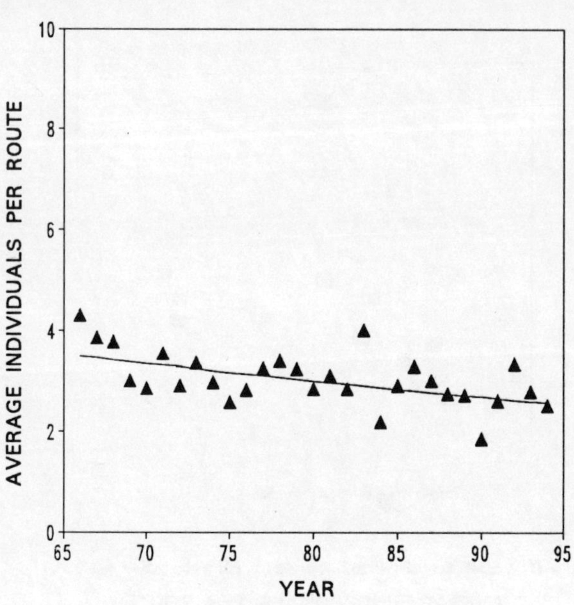

MIGRATORY STATUS: Temperate migrant.

BREEDING HABITAT: Scrub, including shrubby edges and early successional habitat.

ABUNDANCE AND DISTRIBUTION: Fairly common (3.0 birds per route) and widely distributed (all routes). Equally abundant in Western and Eastern Ohio (2.9 vs. 3.1 birds per route, P = 0.76).

OHIO POPULATION TREND: Percent Annual Change (± SE) = -1.1 (± 0.7), P = 0.11
   The annual population decline of 1.1% was not quite statistically significant. Population trends for Western and Eastern Ohio did not differ significantly (-0.7 vs. -1.6%, P = 0.50).

CONTINENTAL AND GREAT LAKES REGIONAL TREND: Similarly, the regional and continental populations have also declined significantly at 1.7 and 1.3% annually.

CEDAR WAXWING

Bombycilla cedrorum

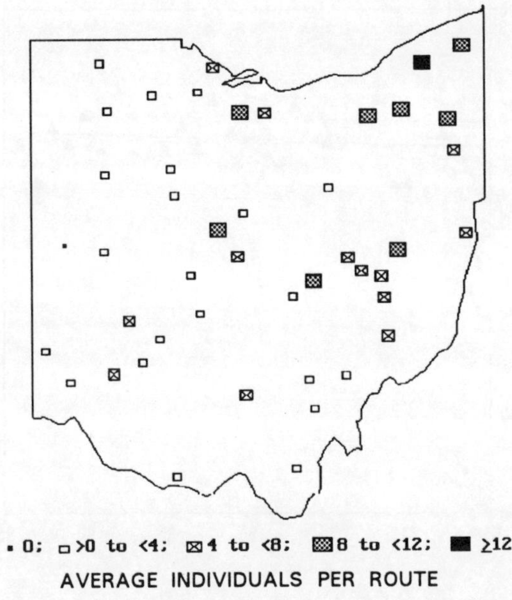

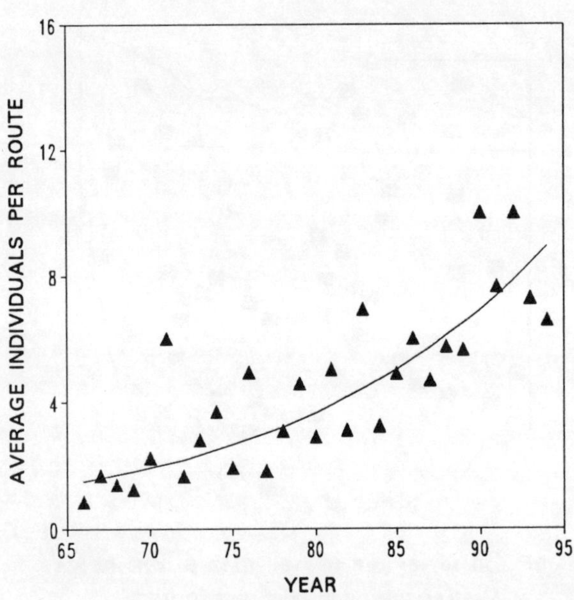

MIGRATORY STATUS: Temperate migrant.

BREEDING HABITAT: Woodlands, including edges and openings. Breed solitarily or in small, loose colonies.

ABUNDANCE AND DISTRIBUTION: Fairly common (4.3 birds per route) and widely distributed (44 routes). Typically a more northerly breeding species within the U.S., Cedar Waxwings were more common in Northern than Southern Ohio (5.8 vs. 3.1 birds per route, P = 0.02). Also more common in Eastern than Western Ohio (6.1 vs. 2.9 birds per route, P = 0.005).

OHIO POPULATION TREND: Percent Annual Change (± SE) = 6.6 (±1.6), P < 0.001
Cedar Waxwings increased significantly at 6.6% annually. The large annual variation is typical of breeding Cedar Waxwings which are reported to be abundant at a given site in some years and absent in others (Peterjohn, 1989; Peterjohn and Rice, 1991).

The population trends for Northern vs. Southern (6.5 vs. 7.2%, P = 0.51) and Eastern vs. Western Ohio (6.2 vs. 7.5%, P = 0.67) did not differ significantly.

CONTINENTAL AND GREAT LAKES REGIONAL TREND: Similarly, the regional and continental populations have also increased significantly at 1.8 and 1.8% annually.

EUROPEAN STARLING

Sturnus vulgaris

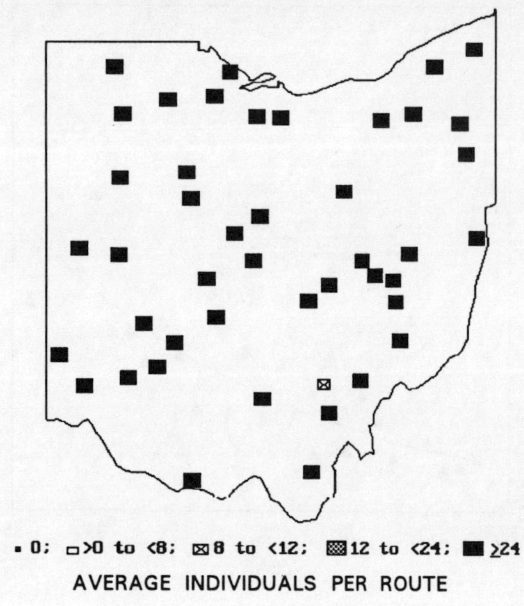

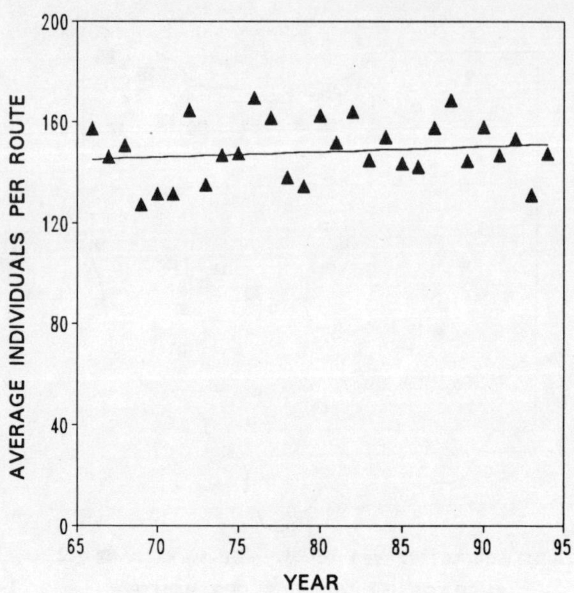

MIGRATORY STATUS: Permanent resident.

BREEDING HABITAT: Residential areas, especially those near agricultural fields.

ABUNDANCE AND DISTRIBUTION: Very abundant (148 birds per route) and widely distributed (all routes). More common in Western Ohio than in Eastern Ohio (189 vs. 102 birds per route, $P < 0.001$).

OHIO POPULATION TREND: Percent Annual Change ($\pm$ SE) = 0.1 ($\pm$ 0.4), P = 0.72
Ohio's European Starling population has not increased or decreased significantly since 1966 (0.1% annual change). The population trends for Western and Eastern Ohio were similar (0.1 vs. 0.2%, P = 0.97).
Starlings invaded Ohio during the 1910s and had become very abundant by the 1930s (Peterjohn, 1989).

CONTINENTAL AND GREAT LAKES REGIONAL TREND: The regional and continental populations have decreased significantly at 0.5 and 1.1% annually.

WHITE-EYED VIREO

Vireo griseus

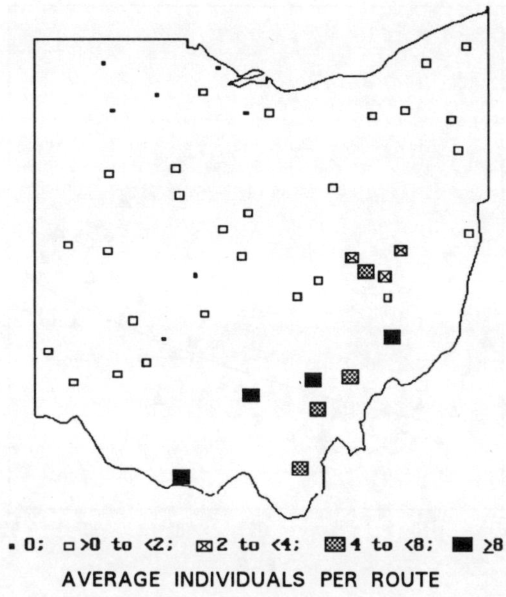

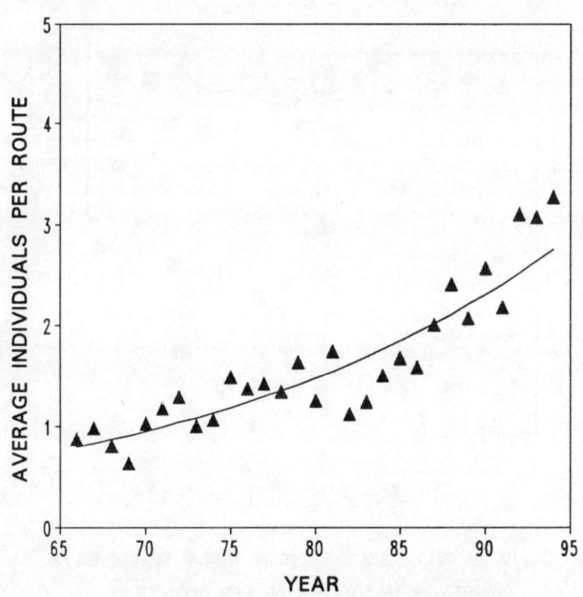

MIGRATORY STATUS: Central neotropical migrant.

BREEDING HABITAT: Scrub, including woodland openings and edges, such as those along streams and roads.

ABUNDANCE AND DISTRIBUTION: Uncommon (1.6 birds per route) and fairly widely distributed (38 routes). More common in the forested Eastern than Western Ohio (2.7 vs. 0.6 birds per route, P = 0.007). Much more common in Southern than Northern Ohio into which they have continued to expand their range (2.7 vs. 0.2, P = 0.001). Especially abundant in the southeastern Unglaciated Plateau (4.0 birds per route).

OHIO POPULATION TREND: Percent Annual Change (± SE) = 4.5 (± 1.2), P = 0.001
   White-eyed Vireos increased significantly at 4.5% annually. White-eyed Vireos increased at a higher rate in Eastern than Western Ohio (5.6 vs. 0.5%, P < 0.001). White-Eyed Vireos are known to have expanded northward within Ohio since the 1930s (Peterjohn, 1989) and this movement is evident in BBS data. White-eyed Vireos were recorded on only 1 BBS route in Northern Ohio prior to 1975 and on 12 routes since 1975. Northern Ohio's population increased at a significantly higher rate than Southern Ohio's (13.0 vs. 3.9%, P < 0.001).

CONTINENTAL AND GREAT LAKES REGIONAL TREND: Neither the regional nor continental population exhibited a significant overall trend (0.9 and 0.0% annually).

YELLOW-THROATED VIREO

Vireo flavifrons

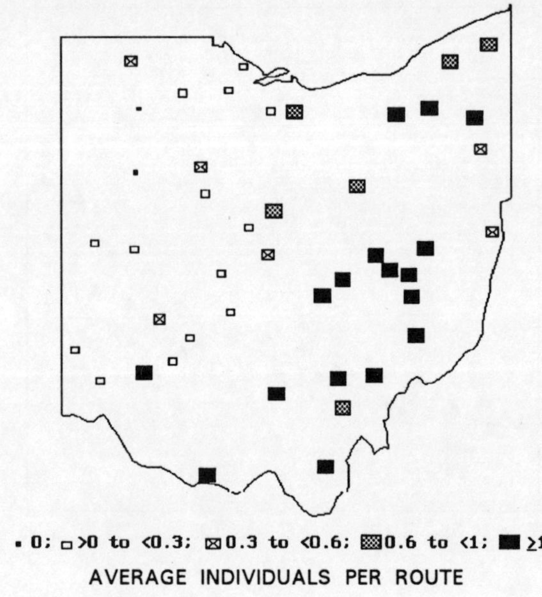

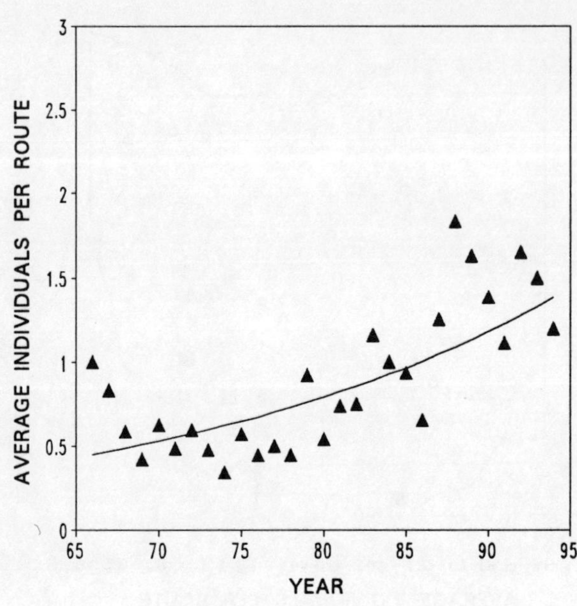

MIGRATORY STATUS: Central neotropical migrant.

BREEDING HABITAT: Mature woodlands.

ABUNDANCE AND DISTRIBUTION: Rare (0.9 birds per route) but widely distributed (43 routes). More common in Eastern than Western Ohio (1.6 vs. 0.3 birds per route, P < 0.001).

OHIO POPULATION TREND: **Percent Annual Change (± SE) = 4.1 (± 1.1), P = 0.001**
   Yellow-throated Vireos increased significantly at 4.1% annually. They increased most sharply after 1978. Increases in Eastern and Western Ohio were similar (4.3 vs. 2.7%, P = 0.45).

CONTINENTAL AND GREAT LAKES REGIONAL TREND: Similarly, the regional and continental populations increased significantly at 2.2 and 1.1% annually.

WARBLING VIREO

Vireo gilvus

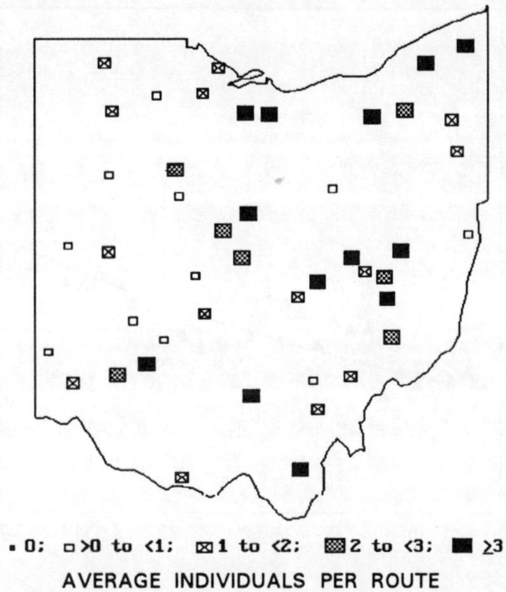

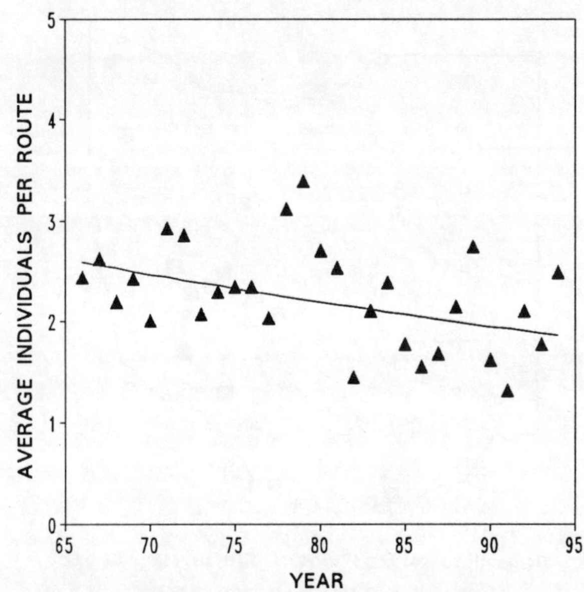

MIGRATORY STATUS: Central neotropical migrant.

BREEDING HABITAT: Mature riparian corridors, especially where tall trees are interspersed with open areas.

ABUNDANCE AND DISTRIBUTION: Uncommon (2.2 birds per route) but widely distributed (all routes). Because Warbling Vireos tend to be in forested riparian areas which are not well sampled by BBS roadside routes, Warbling Vireos are probably more common, relative to other species, than suggested by BBS data.

Warbling Vireos were more common in Eastern than Western Ohio (2.9 vs. 1.6 birds per route, P = 0.03).

OHIO POPULATION TREND: Percent Annual Change ($\pm$ SE) = -1.2 ($\pm$ 0.7), **P = 0.13**

Warbling Vireos declined overall (1.2% annually), but the trend was not significant due to high annual variation in number recorded per route. Warbling Vireos declined significantly in Eastern Ohio at 2.5% annually (P = 0.01), but not in Western Ohio (0.7%, P = 0.47).

CONTINENTAL AND GREAT LAKES REGIONAL TREND: Warbling Vireos in the Great Lakes Region declined significantly at 1.0% annually, however, the continental population increased significantly at 1.1% annually.

RED-EYED VIREO

Vireo olivaceus

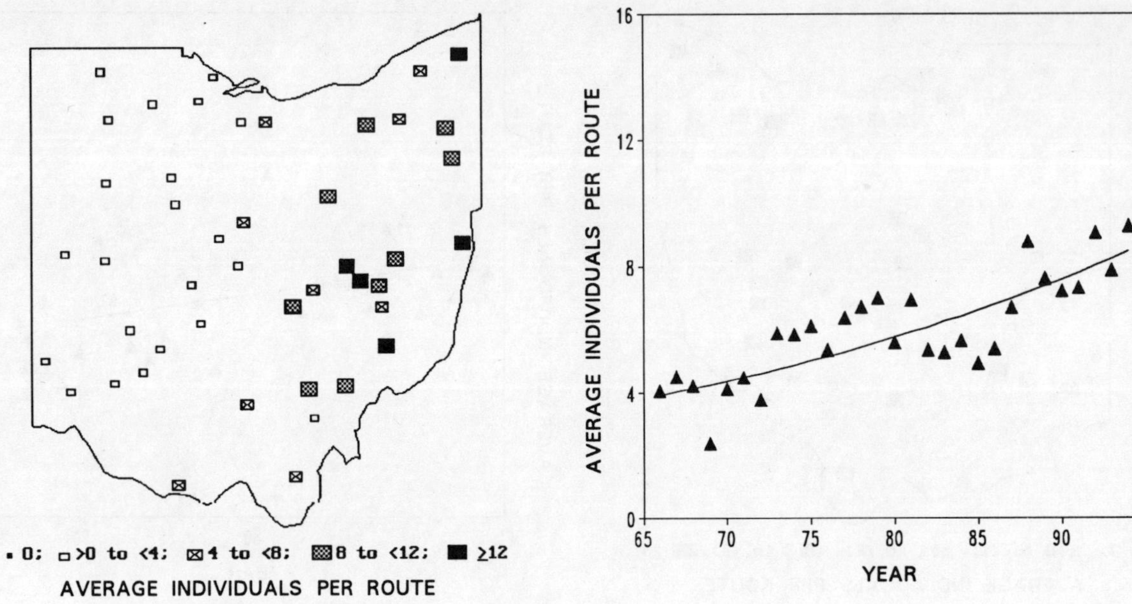

MIGRATORY STATUS: Southern neotropical migrant.

BREEDING HABITAT: Woodlands, including mature and young woodlands.

ABUNDANCE AND DISTRIBUTION: Common (6.1 birds per route) and widely distributed (all routes).
More than 5 times more common in the forests of Eastern than in Western Ohio (10.8 vs. 2.0 birds per route, P
< 0.001).

OHIO POPULATION TREND: Percent Annual Change ($\pm$ SE) = 2.8 ($\pm$ 0.7), P = 0.001
   Red-eyed Vireos increased significantly at 2.8% annually. Trends in Eastern and Western Ohio did not differ
significantly (3.1 vs. 1.9%, P = 0.45).

CONTINENTAL AND GREAT LAKES REGIONAL TREND: The regional and continental populations
also increased significantly at 1.3 and 1.0% annually.

BLUE-WINGED WARBLER

Vermivora pinus

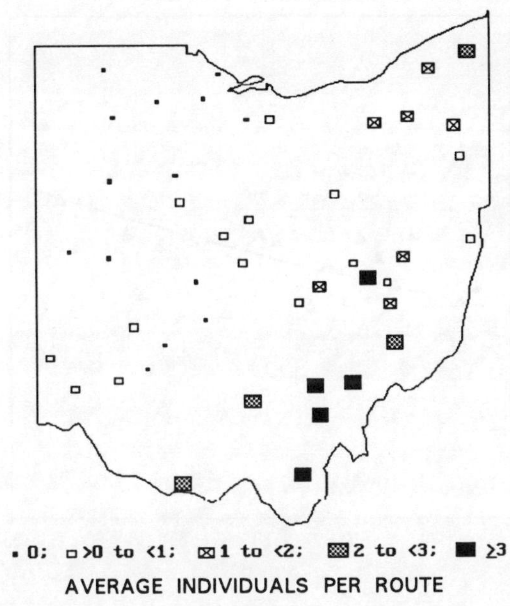

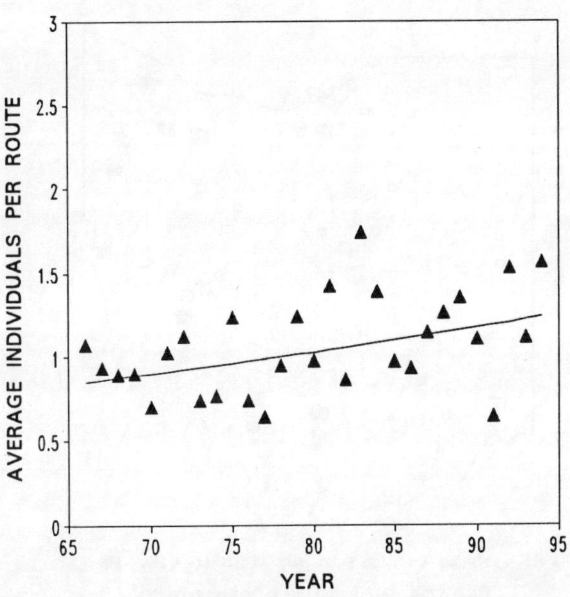

MIGRATORY STATUS: Central neotropical migrant.

BREEDING HABITAT: Scrub, including edge and early successional habitat.

ABUNDANCE AND DISTRIBUTION: Rare (1.0 birds per route) and fairly widely distributed (31 routes).
Abundance >10 times higher in Eastern than Western Ohio (2.0 vs. 0.14 birds per route, P < 0.001) and
somewhat higher in Southern than Northern Ohio (1.4 vs. 0.5 birds per route, P = 0.049).

OHIO POPULATION TREND: Percent Annual Change (± SE) = 1.3 (± 0.9), P = 0.17
  Blue-winged Warblers did not exhibit a significant 1966-94 trend (+1.3% annual change). Trends in Eastern
and Western Ohio did not differ significantly (1.5 vs. -0.2%, P = 0.10), and trends in Southern and Northern
Ohio did not differ significantly (0.9 vs. 2.8%, P = 0.50).

CONTINENTAL AND GREAT LAKES REGIONAL TREND: Similarly, the regional and continental
populations did not exhibit significant 1966-94 trends (2.1 and 0.4% annual change).

YELLOW WARBLER

Dendroica petechia

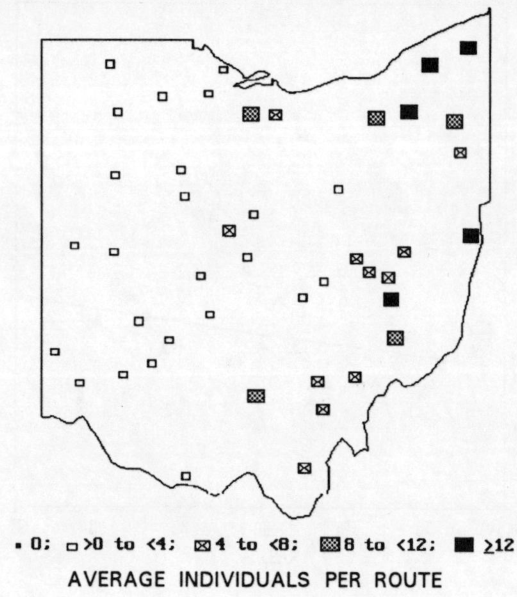

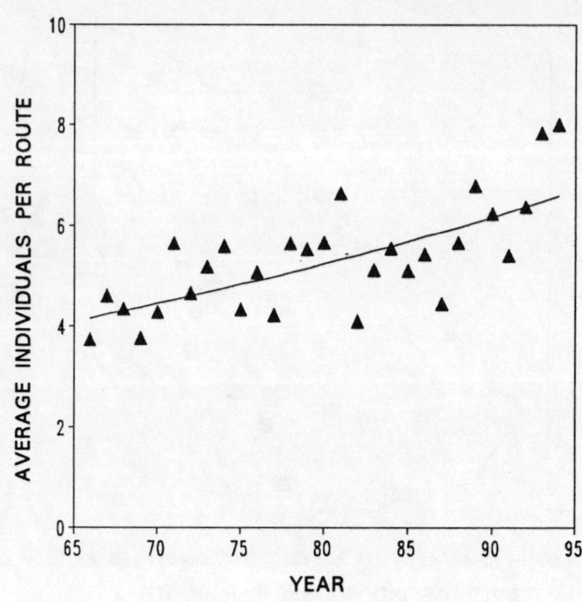

MIGRATORY STATUS: Central neotropical migrant.

BREEDING HABITAT: Scrub, particularly thickets bordering wetlands.

ABUNDANCE AND DISTRIBUTION: Common (5.2 birds per route) and widely distributed (all routes). Much more common in Eastern than Western Ohio (8.9 vs. 1.9 birds per route, P < 0.001).

OHIO POPULATION TREND: Percent Annual Change (± SE) = 1.7 (± 1.0), P = 0.11
   Yellow Warblers appear to be increasing, but large annual variation masks the overall trend (1.7% annual change). Yellow Warblers increased significantly at 3.6% annually in Western Ohio (P = 0.03), where they are less common, but showed substantial variation and no significant trend in Eastern Ohio (1.3%, P = 0.26).

CONTINENTAL AND GREAT LAKES REGIONAL TREND: The regional and continental populations increased significantly at 1.4 and 0.6% annually.

YELLOW-THROATED WARBLER

Dendroica dominica

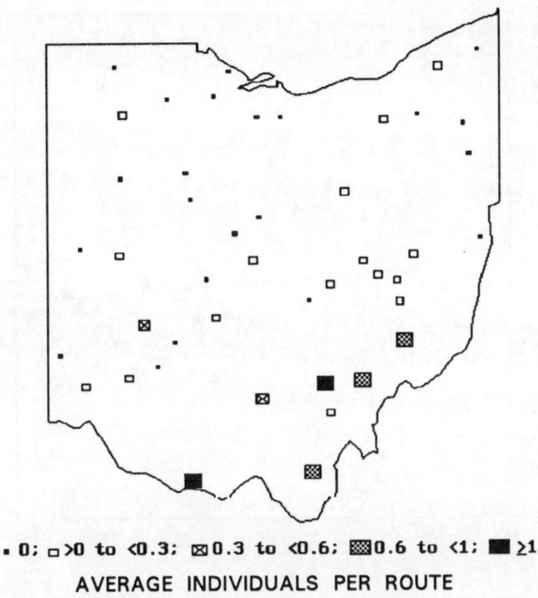

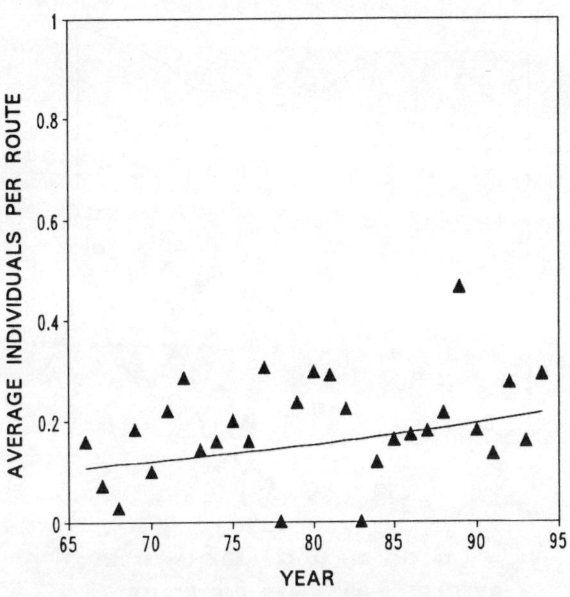

MIGRATORY STATUS: Central neotropical migrant.

BREEDING HABITAT: Mature woodlands, particularly riparian corridors.

ABUNDANCE AND DISTRIBUTION: Rarely recorded (0.18 birds per route) and locally distributed (23 routes). Because riparian corridors in forest interiors are poorly sampled by roadside BBS routes, Yellow-throated Warblers are probably more common, relative to other songbirds, than suggested by BBS data.

More common in Southern than Northern Ohio (0.31 vs. 0.02 birds per route, P = 0.006). Equally rare in Eastern and Western Ohio (0.24 vs. 0.12 birds per route, P = 0.30).

OHIO POPULATION TREND: Percent Annual Change (± SE) = 2.5 (± 0.9), P = 0.008

Yellow-throated Warblers have increased significantly at 2.5% annually since 1966; the overall trend is significant despite substantial annual variation. Inadequate sample to compare regional trends within Ohio.

Historically, Yellow-throated Warblers, which were once common throughout Ohio, declined sharply during the late 1800s (Peterjohn, 1989). The population slowly spread northward into its former range, but Yellow-throated Warblers are still more common in southern Ohio.

CONTINENTAL AND GREAT LAKES REGIONAL TREND: Neither the regional nor continental population exhibited a significant overall trend (2.2 and 0.8% annual change).

PRAIRIE WARBLER

Dendroica discolor

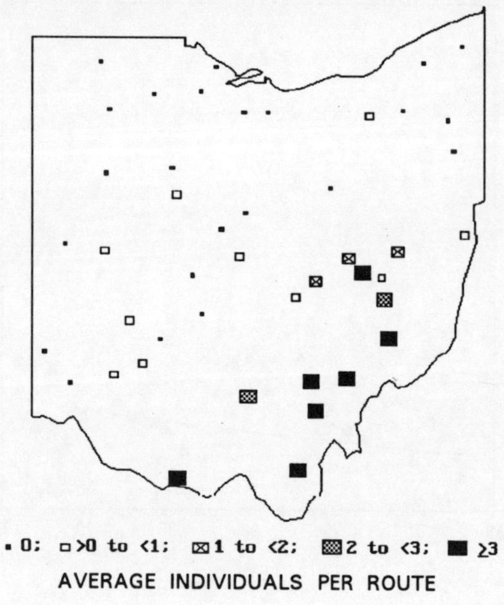

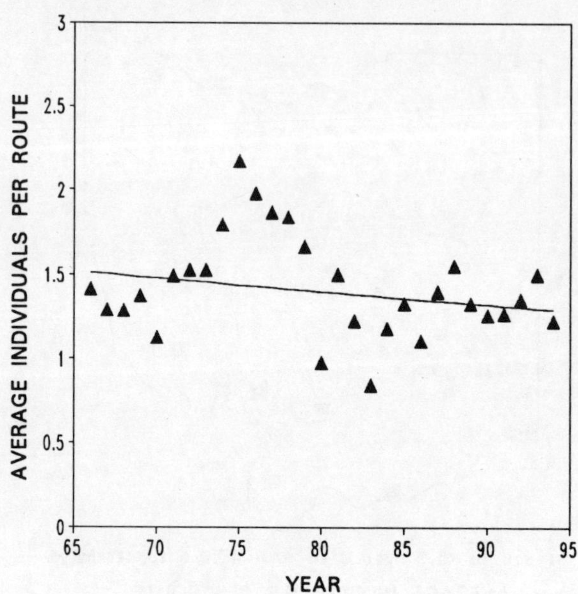

MIGRATORY STATUS: Central neotropical migrant.

BREEDING HABITAT: Scrub, especially abandoned fields that have reverted to the shrub/sapling stage successional habitat.

ABUNDANCE AND DISTRIBUTION: Uncommon (1.3 birds per route) and locally distributed (22 routes). Typical breeding range includes only southern Ohio, and Prairie Warblers are significantly more common on BBS routes in Southern than Northern Ohio (2.4 vs. 0.1, P = 0.01).

OHIO POPULATION TREND: Percent Annual Change (± SE) = -0.6 (± 0.7), P = 0.41

Ohio's Prairie Warbler population experienced particularly low numbers around 1970 and in the early 1980s, but the overall 1966-94 trend is not statistically significant (-0.6% annual change). Prairie Warblers are thought to have been expanding northward in Ohio until the late 1970s (Peterjohn and Rice, 1991), and the initial 1966-75 increase may have reflected this expansion; declines in subsequent years may reflect loss of early successional habitat (see also Robbins et al., 1986).

Prairie Warblers occurred on too few routes in Western and Northern Ohio to allow comparisons of trends between regions.

CONTINENTAL AND GREAT LAKES REGIONAL TREND: Both the regional and continental populations have declined fairly dramatically at significant annual rates of 2.0 and 2.7%. The decline is thought to be due to loss of the preferred scrub growth habitat (Robbins et al., 1986).

CERULEAN WARBLER

Dendroica cerulea

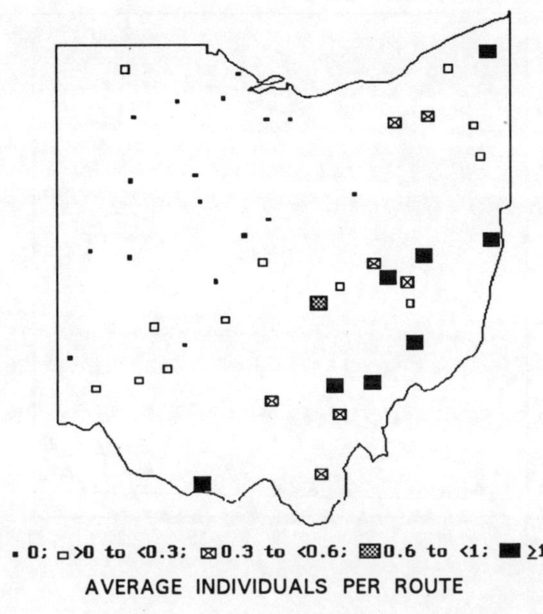

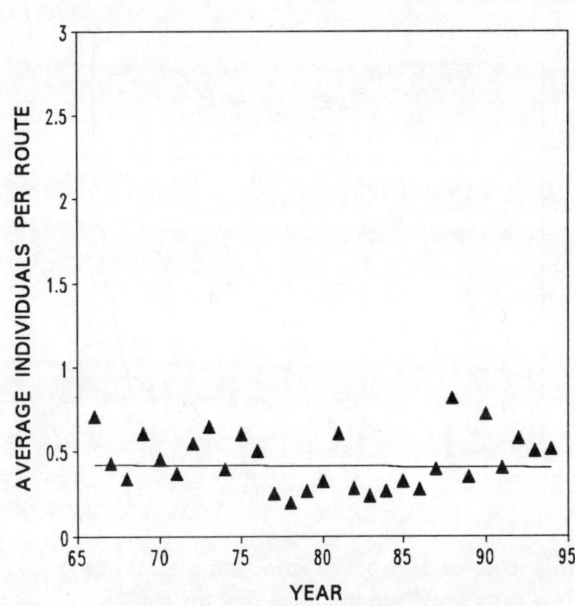

MIGRATORY STATUS: Southern neotropical migrant.

BREEDING HABITAT: Mature woodlands, especially large tracts.

ABUNDANCE AND DISTRIBUTION: Rarely recorded (0.4 birds per route) and locally distributed (28 routes). Because roadside BBS routes do not adequately sample the interiors of large woodlands, Cerulean Warblers are probably relatively more common than indicated by BBS data.

More common in the forests of Eastern Ohio than in Western Ohio (0.8 vs. 0.1, P < 0.001).

OHIO POPULATION TREND: Percent Annual Change ($\pm$ SE) = -0.1 ($\pm$ 1.2), P = 0.91

Ohio's Cerulean Warbler population has not exhibited a significant overall trend (-0.1% annual change). Cerulean Warblers occurred on too few routes in Western Ohio to allow a comparison of Eastern vs. Western trends.

CONTINENTAL AND GREAT LAKES REGIONAL TREND: The regional and continental populations have declined dramatically and significantly at 4.9 and 4.3% annually.

BLACK-AND-WHITE WARBLER

Mniotilta varia

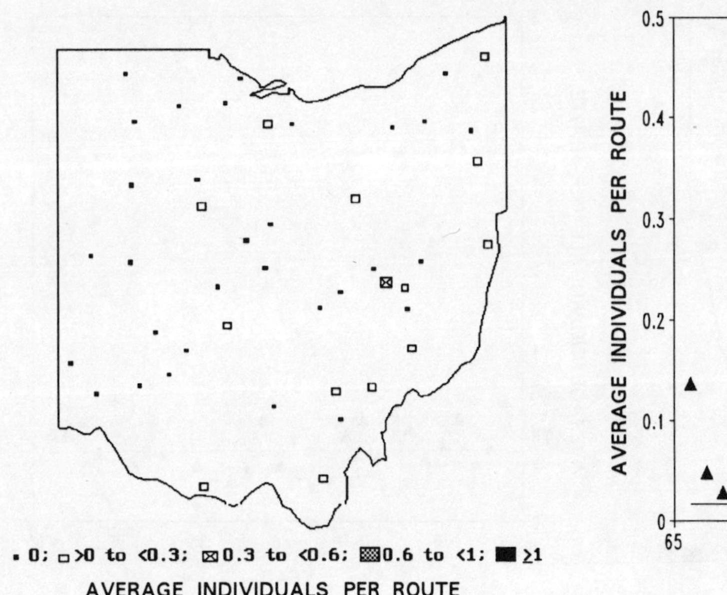

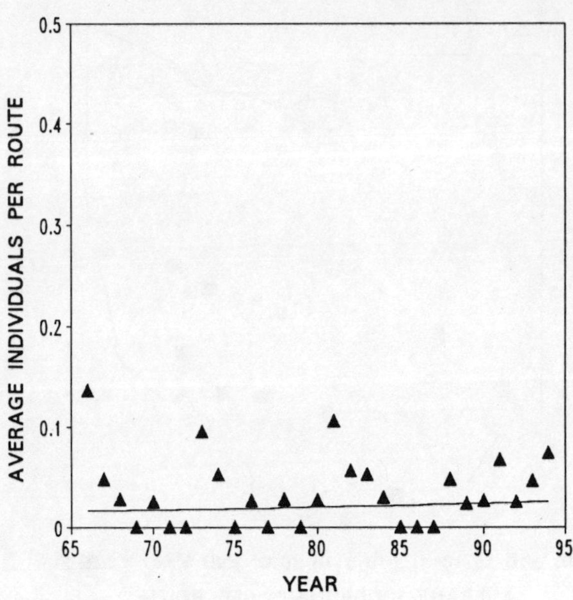

MIGRATORY STATUS: Central neotropical migrant.

BREEDING HABITAT: Prefer extensive young woodlands, especially those with dense understories.

ABUNDANCE AND DISTRIBUTION: Rare (0.04 birds per route) and very locally distributed (14 routes). More common in Eastern than Western Ohio (0.07 vs. 0.008, P = 0.04).

OHIO POPULATION TREND: **Percent Annual Change ($\pm$ SE) = 1.7 ($\pm$ 3.0), P = 0.58**
   Black-and-White Warblers did not exhibit a signficant overall trend. The low abundance, local distribution, and large year-to-year variation in number recorded suggest that better data are needed and that BBS data for Ohio's Black-and-White Warbler population should be interpreted with caution.
   Data were insufficient to compare population trends in Eastern vs. Western Ohio.

CONTINENTAL AND GREAT LAKES REGIONAL TREND: Similarly, Black-and-White Warblers did not exhibit a significant regional or continental population trend (both 0.5).

AMERICAN REDSTART

Setophaga ruticilla

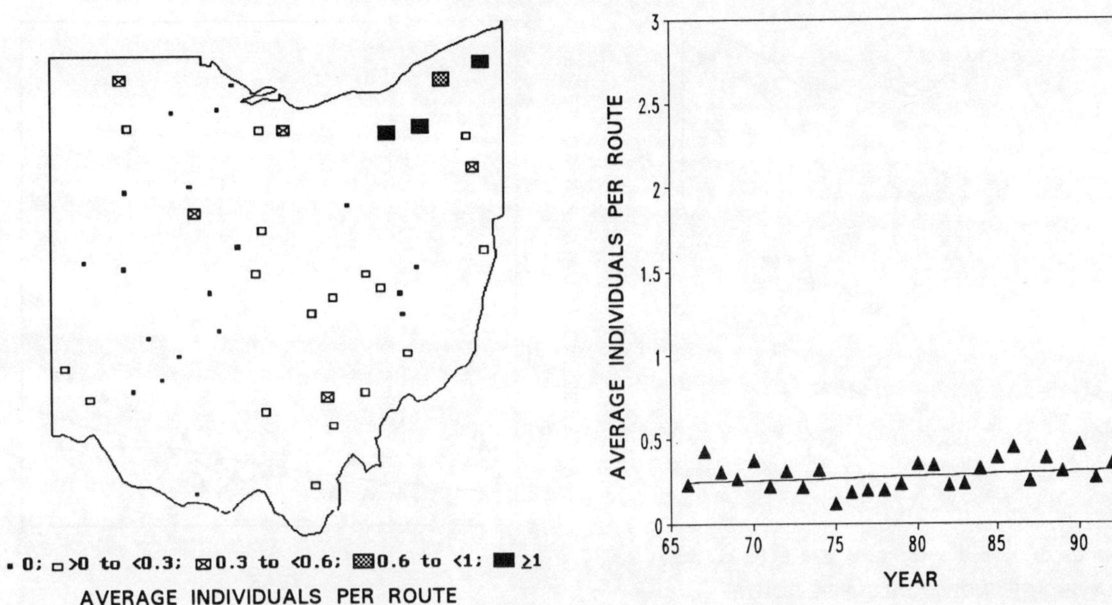

MIGRATORY STATUS: Central neotropical migrant.

BREEDING HABITAT: Young woodlands, especially those with dense understories.

ABUNDANCE AND DISTRIBUTION: Rare (0.3 birds per route) and locally distributed (26 routes). Tend to be more abundant in Eastern than Western Ohio, but the difference is not significant (0.48 vs. 0.08, P = 0.10). Especially abundant in the northeastern Glaciated Plateau (1.2 birds per route).

OHIO POPULATION TREND: Percent Annual Change (± SE) = 0.9 (± 1.5), P = 0.54
   American Redstarts in Ohio have remained relatively stable (0.9% annual change). American Redstarts appear to have increased in Western Ohio (4.3%, P = 0.03) and to have shown high annual variation in Eastern Ohio (0.6%, p = 0.70), although each trend is based on only 9-14 routes.

CONTINENTAL AND GREAT LAKES REGIONAL TREND: Likewise, the regional and continental populations have not exhibited significant overall trends (-0.5 and -0.5% annual change).

OVENBIRD

Seiurus aurocapillus

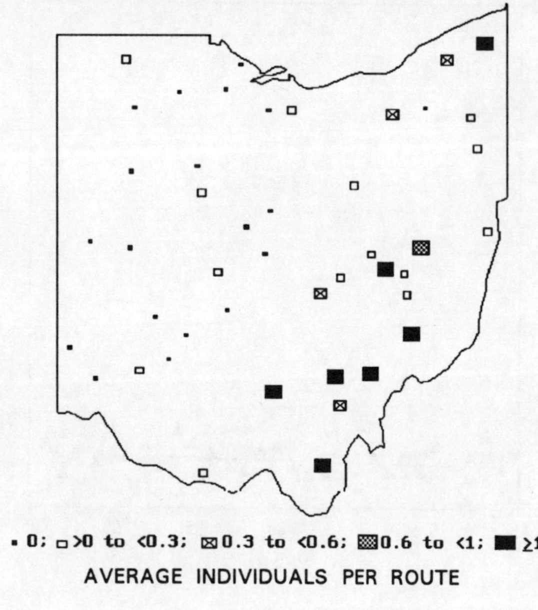

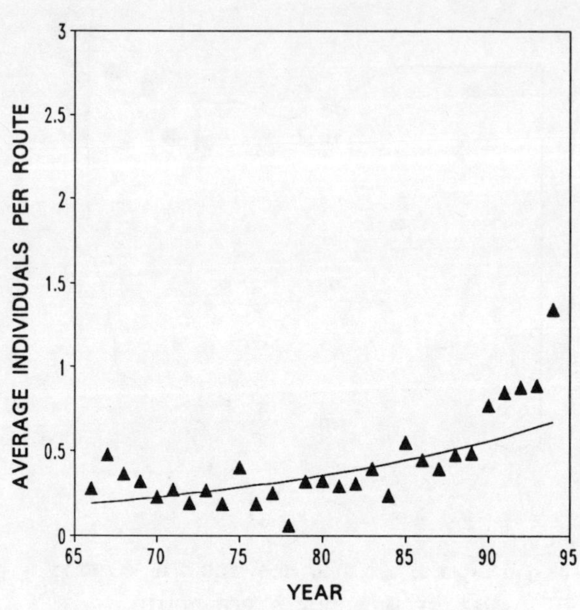

MIGRATORY STATUS: Central neotropical migrant.

BREEDING HABITAT: Mature woodlands, especially interiors of fairly open extensive woodlands.

ABUNDANCE AND DISTRIBUTION: Rarely recorded (0.4 birds per route) and locally distributed (26 routes). Like other forest interior species, Ovenbirds are underrepresented, relative to other species, in BBS data.

Much more common in the forests of Eastern than in Western Ohio (0.86 vs. 0.03, P = 0.002).

OHIO POPULATION TREND: Percent Annual Change ($\pm$ SE) = 4.6 ($\pm$ 1.4), P = 0.003

Ovenbirds increased significantly, at 4.6% annually. The increase was particularly apparent in the last decade. Too rare in Western Ohio to allow a comparison of Eastern vs. Western trend.

CONTINENTAL AND GREAT LAKES REGIONAL TREND: Similarly, the regional and continental populations have increased significantly at 1.0 and 0.6% annually.

LOUISIANA WATERTHRUSH

Seiurus motacilla

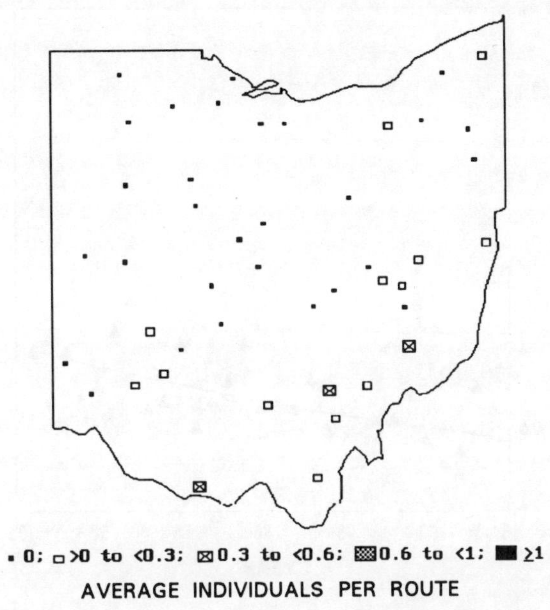

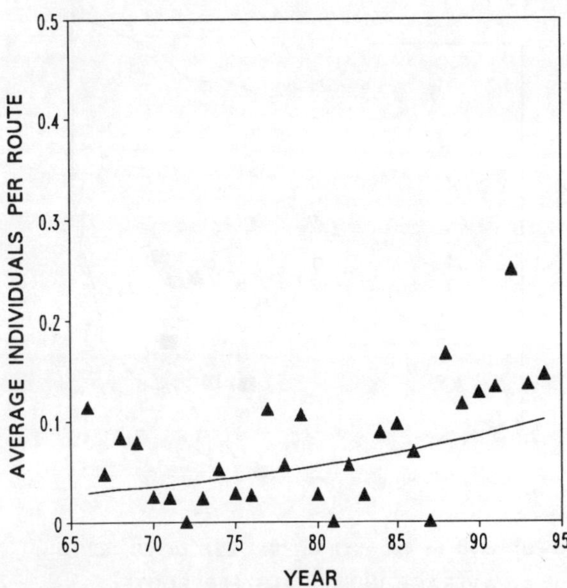

MIGRATORY STATUS: Central neotropical migrant.

BREEDING HABITAT: Mature woodlands, especially along small, fast-flowing streams.

ABUNDANCE AND DISTRIBUTION: Rare (0.07 birds per route) and very locally distributed (16 routes). Because roadside BBS routes do not adequately sample forest interior riparian corridors, Louisiana Waterthrushes are probably relatively more common than indicated by BBS data.

Somewhat more common in Eastern than Western Ohio (0.12 vs. 0.03, P = 0.06), and significantly more common in the Unglaciated Plateau of southeastern Ohio than elsewhere (0.18 vs. 0.03, P = 0.01).

OHIO POPULATION TREND: Percent Annual Change (± SE) = 4.5 (± 2.5), P = 0.07

Louisiana Waterthrushes increased at 4.5% annually, but the trend was not quite statistically significant due to the large year-to-year variation in number recorded. Too rare in Western Ohio to allow a comparison of Eastern and Western trends.

CONTINENTAL AND GREAT LAKES REGIONAL TREND: The regional population has increased at a high and significant rate of 5.3% annually; however, the continental population did not exhibit a significant overall trend (0.4% annual change).

KENTUCKY WARBLER

Oporornis formosus

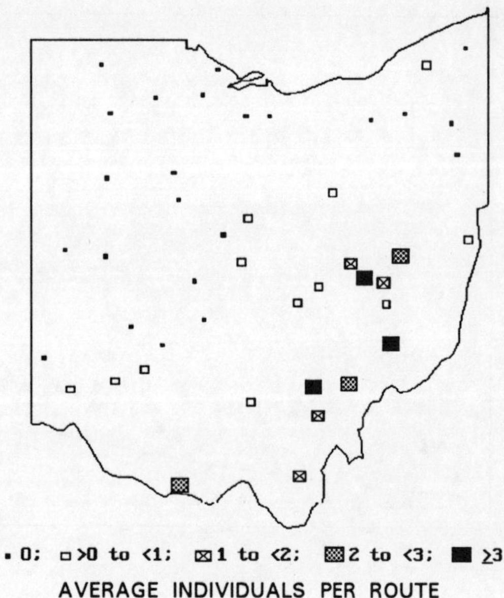

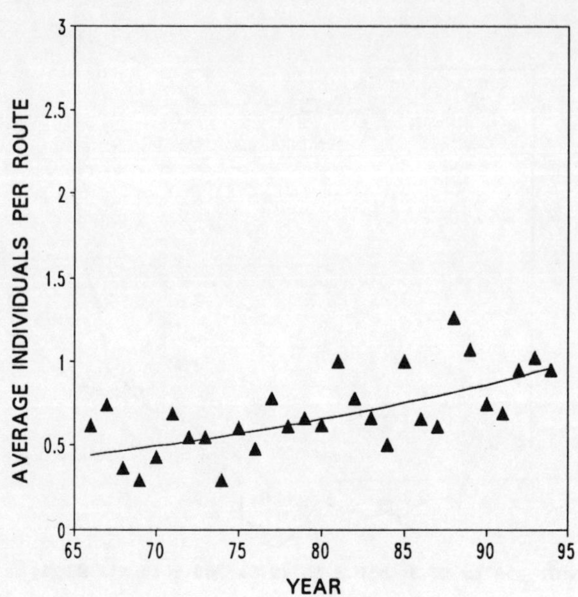

MIGRATORY STATUS: Central neotropical migrant.

BREEDING HABITAT: Mature woodlands, especially forest interiors.

ABUNDANCE AND DISTRIBUTION: Rarely recorded on BBS routes (0.7 birds per route) and locally distributed (22 routes). Like other forest interior species, Kentucky Warblers are probably relatively more common than indicated by roadside BBS routes.

More than 10 times more common in Eastern than Western Ohio (1.3 vs. 0.1, P = 0.003). Kentucky Warblers have spread northward during the 20th century and are still more rare in Northern than Southern Ohio (1.2 vs. 0.2, P = 0.006).

OHIO POPULATION TREND: Percent Annual Change (± SE) = 2.8 (± 1.1), P = 0.02

Kentucky Warblers increased at a significant annual rate of 2.8%. Too rare in Western Ohio to allow a comparison of Eastern and Western trends.

CONTINENTAL AND GREAT LAKES REGIONAL TREND: The regional population increased somewhat, but nonsignificantly (1.3% annually), and the continental population decreased significantly at 1.0% annually.

COMMON YELLOWTHROAT

Geothlypis trichas

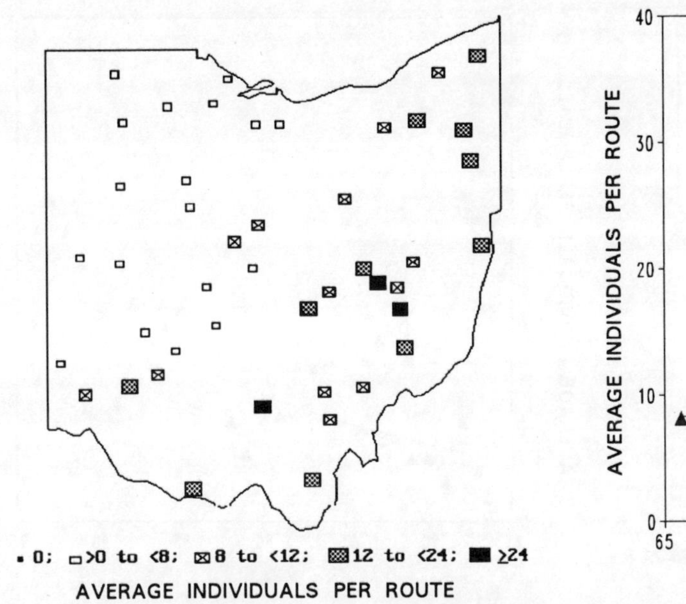

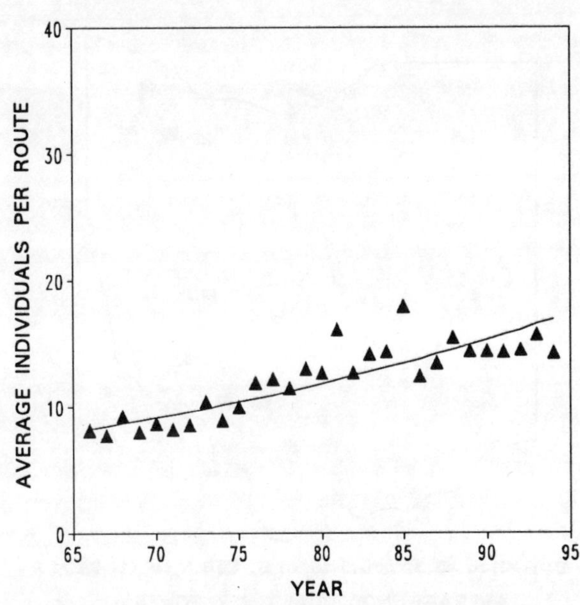

MIGRATORY STATUS: Central neotropical migrant.

BREEDING HABITAT: Scrub, especially dense thickets near water.

ABUNDANCE AND DISTRIBUTION: Common (11.9 birds per route) and widely distributed (all routes). More than twice as common in Eastern than Western Ohio (18.2 vs. 6.5 birds per route, P < 0.001).

OHIO POPULATION TREND: Percent Annual Change (± SE) = 2.6 (± 0.7), P = 0.001
   Common Yellowthroats have increased significantly and steadily since 1966 (2.6% annual change). Increases in Eastern and Western Ohio were similar (2.4 vs. 3.1% annually, P = 0.60).

CONTINENTAL AND GREAT LAKES REGIONAL TREND: In contrast, the regional population did not change significantly (0.0% annually), and the continental population declined significantly at -0.4% annually.

HOODED WARBLER

Wilsonia citrina

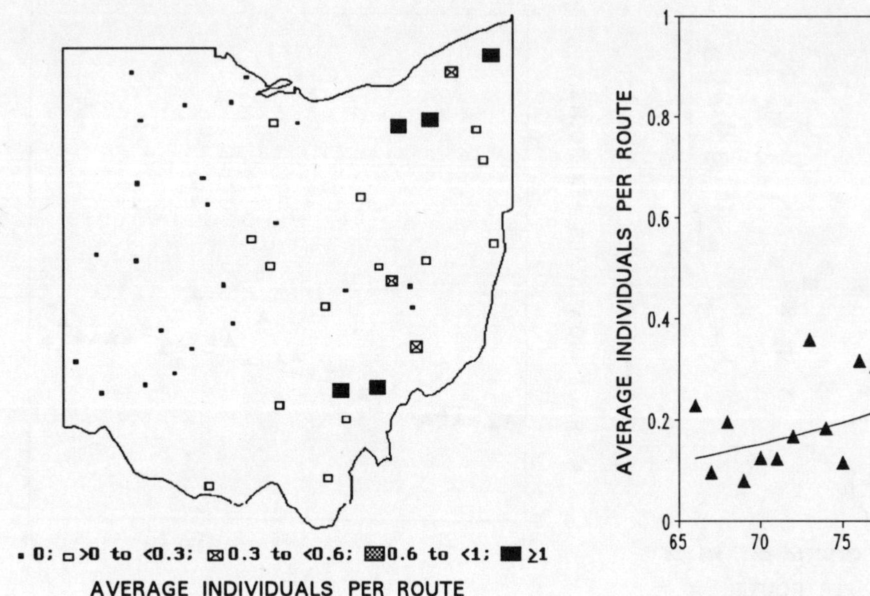

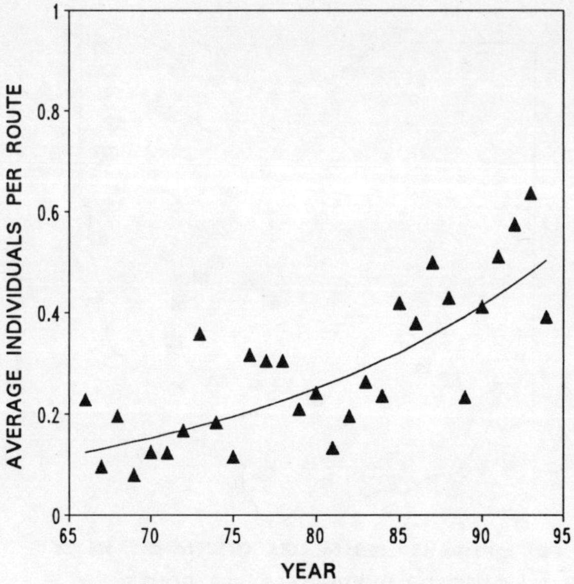

MIGRATORY STATUS: Central neotropical migrant.

BREEDING HABITAT: Mature woodlands, especially near openings in the forest interior where the understory is well-developed.

ABUNDANCE AND DISTRIBUTION: Rarely recorded (0.3 birds per route) and locally distributed (22 routes). Like other forest interior species, Hooded Warblers are underrepresented, relative to other species, by BBS data.

Much more common in the forests of Eastern Ohio than in Western Ohio where it occurred on only 4 routes (0.55 vs. 0.01, P = 0.003).

OHIO POPULATION TREND: Percent Annual Change (± SE) = 5.1 (± 2.0), P = 0.02

Hooded Warblers exhibited a significantly increasing trend of 5.1% annually, despite large annual variation in number recorded per route. Hooded Warblers have expanded their range into west-central Ohio since the 1930s (Peterjohn, 1989). Too rare in Western Ohio to allow a comparison of Eastern and Western trends.

CONTINENTAL AND GREAT LAKES REGIONAL TREND: As in Ohio, Hooded Warblers in the Great Lakes Region increased significantly at 5.6% annually. The continental population exhibited no significant trend (0.4% annual change).

YELLOW-BREASTED CHAT

Icteria virens

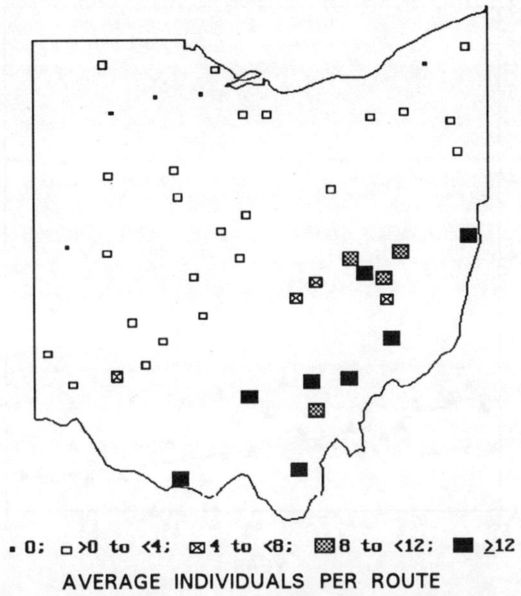

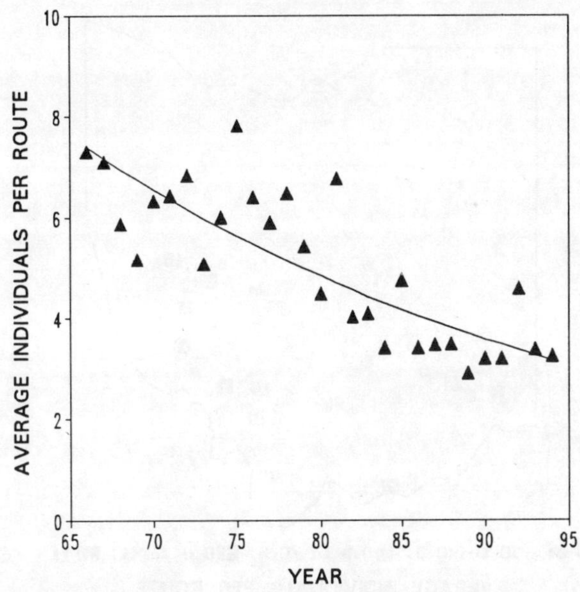

MIGRATORY STATUS: Central neotropical migrant.

BREEDING HABITAT: Scrub, especially abandoned fields dominated by shrubs and saplings.

ABUNDANCE AND DISTRIBUTION: Common (4.9 birds per route) and widely distributed (40 routes). Much more common in the southeastern quarter of the state (Unglaciated Plateau) than elsewhere (12.4 vs. 1.4 birds per route, P < 0.001). Relative scarcity in the northern counties probably explained by Ohio's location at the northern edge of the breeding range (Peterjohn, 1989).

OHIO POPULATION TREND: Percent Annual Change (± SE) = -3.0 (± 0.4), P < 0.001
   Yellow-breasted Chat decreased significantly and dramatically at 3.0% annually. The decrease was significantly more pronounced in Eastern than Western Ohio (4.1 vs. 2.7%, P = 0.02) and Southern than Northern Ohio (3.4 vs. 1.6%, P = 0.03).

CONTINENTAL AND GREAT LAKES REGIONAL TREND: As in Ohio, Yellow-breasted Chat in the regional and continental population decreased significantly at 2.4 and 0.5% annually.

SUMMER TANAGER

Piranga rubra

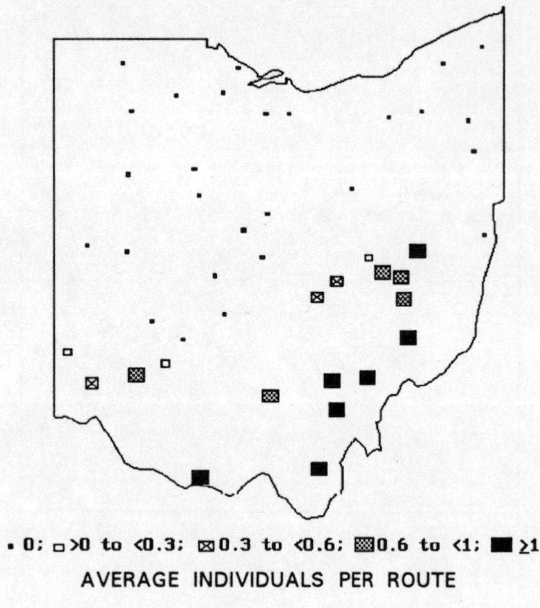

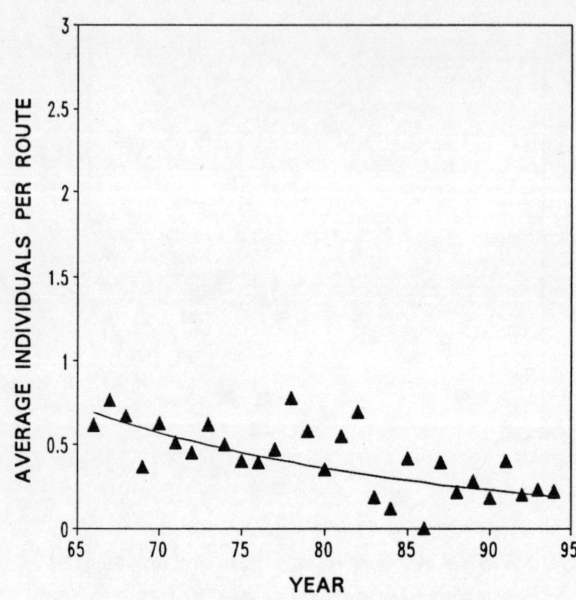

MIGRATORY STATUS: Central neotropical migrant.

BREEDING HABITAT: Mature woodlands.

ABUNDANCE AND DISTRIBUTION: Within Ohio, the Summer Tanager's typical breeding range includes only Southern Ohio (Peterjohn and Rice, 1991) where they are rare (0.7 birds per route) and very locally distributed (18 routes).

OHIO POPULATION TREND: Percent Annual Change ($\pm$ SE) = -4.5 ($\pm$ 1.7), P = 0.02

Summer Tanagers have declined at a striking and significant rate of 4.5% annually. Too rare in Western Ohio to allow a comparison of Eastern and Western trends.

CONTINENTAL AND GREAT LAKES REGIONAL TREND: Neither the regional nor continental population exhibited a significant overall trend (-0.2 and -0.2% annually).

SCARLET TANAGER

Piranga olivacea

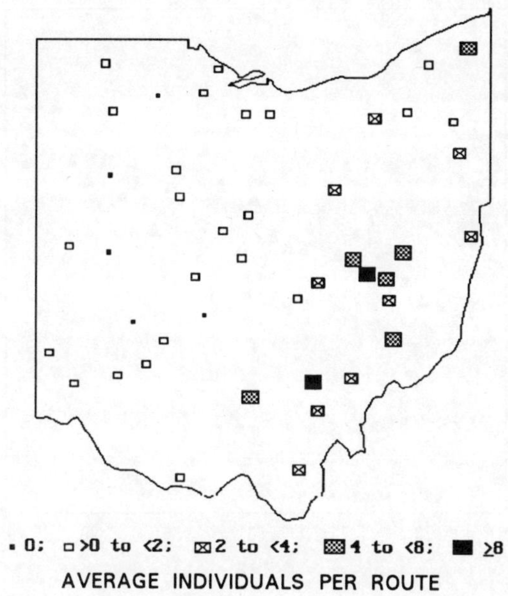

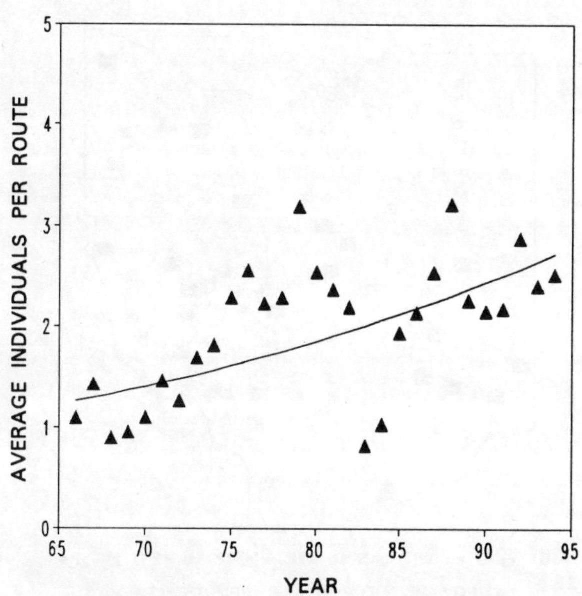

MIGRATORY STATUS: Southern neotropical migrant.

BREEDING HABITAT: Mature woodlands.

ABUNDANCE AND DISTRIBUTION: Fairly common (2.0 birds per route) and widely distributed (40 routes). More than 10 times more common in the forests of Eastern Ohio than in Western Ohio (3.9 vs. 0.3, P < 0.001).

OHIO POPULATION TREND: Percent Annual Change (± SE) = 2.8 (± 0.6), P < 0.001

Scarlet Tanagers increased significantly at 2.8% annually. The trend remains significant if the apparent outliers in 1983 and 1984 are removed. Scarlet Tanagers increased significantly in both Eastern (2.6%, P = 0.001) and Western Ohio (5.6%, P = 0.01), and the two trends did not differ significantly from one another (P = 0.18).

It is surprising that Scarlet Tanagers have not declined in western counties due to intensive agriculture (e.g., Peterjohn, 1989). The increase in eastern counties probably reflects the maturation of previously harvested forests (Peterjohn, 1989).

CONTINENTAL AND GREAT LAKES REGIONAL TREND: Like Ohio's population, Scarlet Tanagers in the Great Lakes Region increased significantly at 1.4% annually. The continental population did not exhibit a significant overall trend (0.0% annual change).

NORTHERN CARDINAL

Cardinalis cardinalis

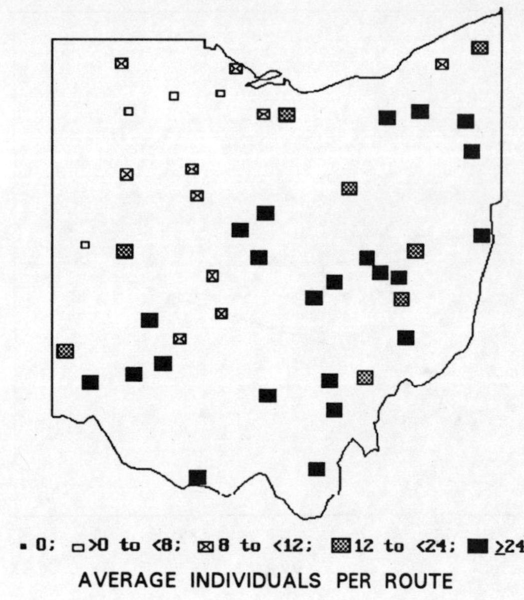

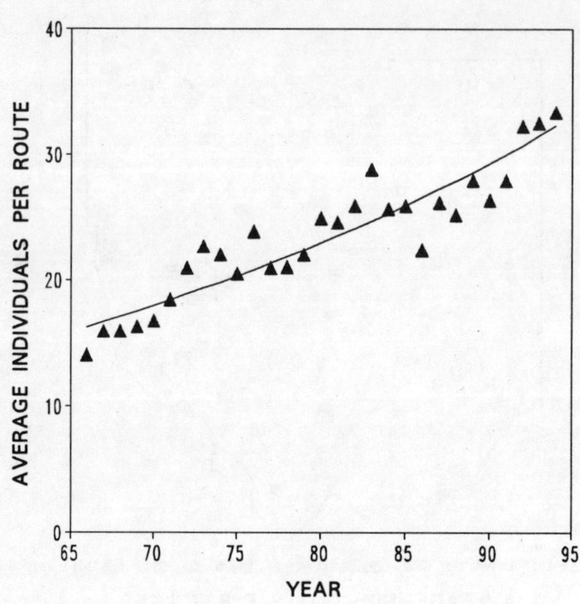

MIGRATORY STATUS: Permanent resident.

BREEDING HABITAT: Scrub, including dense thickets in abandoned fields, along fencerows and woodland edges, and around residential areas.

ABUNDANCE AND DISTRIBUTION: Abundant (23.3 birds per route) and widely distributed (all routes). More common in Eastern than Western Ohio (27.0 vs. 20.1 birds per route, P = 0.04).

Historically, Northern Cardinals probably increased in Southern Ohio as early settlers cleared forests, and had spread northward throughout the state by 1900 (Peterjohn, 1989).

OHIO POPULATION TREND: Percent Annual Change (± SE) = 2.5 (± 0.5), P < 0.001

Northern Cardinals increased significantly and fairly steadily at 2.5% annually. The rate of increase in Eastern and Western Ohio did not differ significantly (1.8 vs. 3.2%, P = 0.14).

CONTINENTAL AND GREAT LAKES REGIONAL TREND: As in Ohio, Northern Cardinals in the Great Lakes Region increased significantly at 0.7% annually, however, the continental population did not exhibit a significant overall trend (-0.1% annual change).

ROSE-BREASTED GROSBEAK

Pheucticus ludovicianus

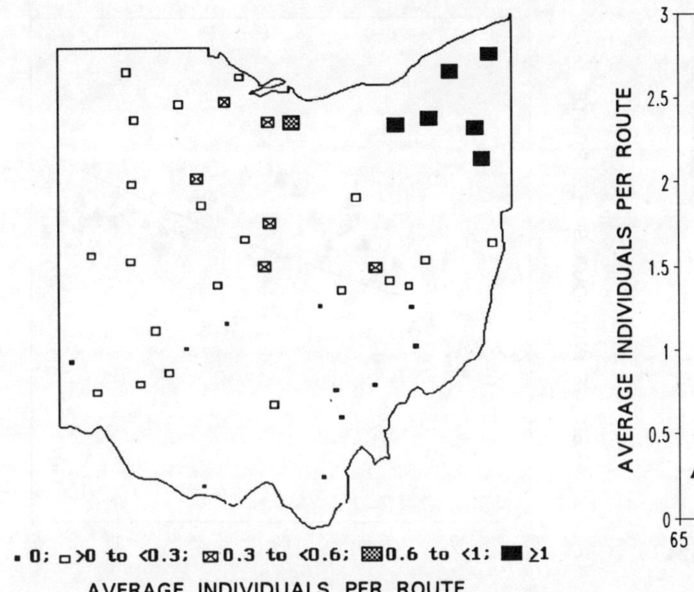

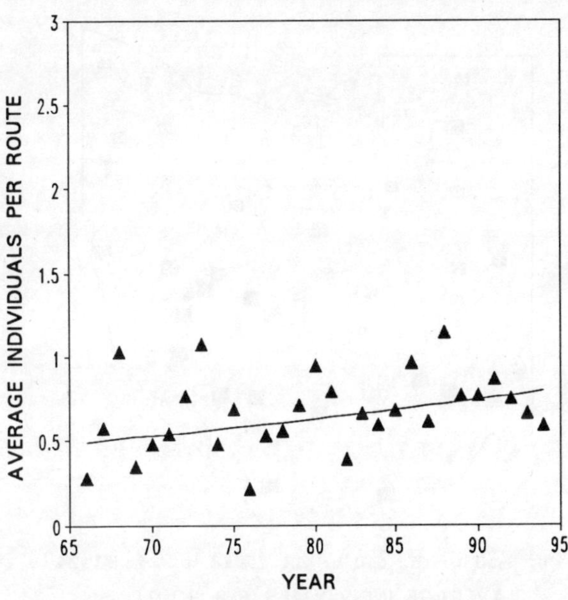

MIGRATORY STATUS: Southern neotropical migrant.

BREEDING HABITAT: Woodlands, including young and mature woodlands, corridors and edges.

ABUNDANCE AND DISTRIBUTION: Rare (0.6 birds per route) and fairly widely distributed (34 routes). Ohio is on the southern edge of their breeding range, and Rose-breasted Grosbeaks are much more common in the northeast corner (Glaciated Plateau) than elsewhere in Ohio (3.3 vs. 0.14 birds per route, P = 0.003). Their range has expanded south and westward since the 1950s (Peterjohn, 1989).

OHIO POPULATION TREND: Percent Annual Change (± SE) = 1.7 (± 1.9), P = 0.37
 Ohio's Rose-breasted Grosbeak population exhibited large annual variation in number recorded per route, due in part to range expansion, and did not exhibit a significant overall trend (1.7% annual change). Their southward and westward range expansion is reflected in a significantly increasing annual change in Southern Ohio (7.5%, P < 0.001) and Western Ohio (7.4%, P = 0.002) but not in either Northern (1.7%, P = 0.41) or Eastern Ohio (1.2%, P = 0.58).

CONTINENTAL AND GREAT LAKES REGIONAL TREND: Similarly, neither the regional nor continental population exhibited a significant overall trend (0.0 and -0.5% annual change).

INDIGO BUNTING

Passerina cyanea

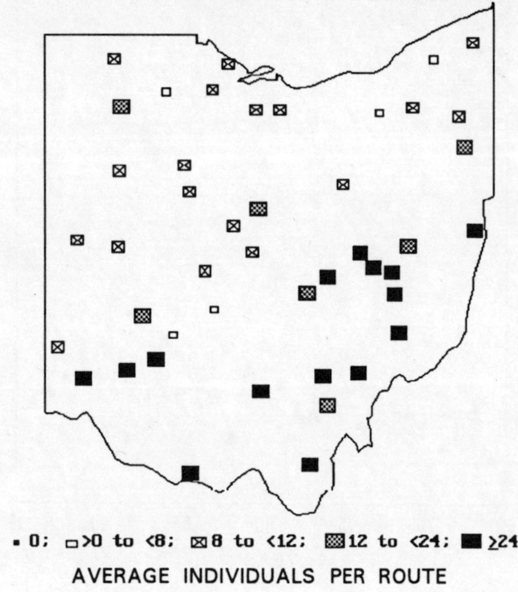

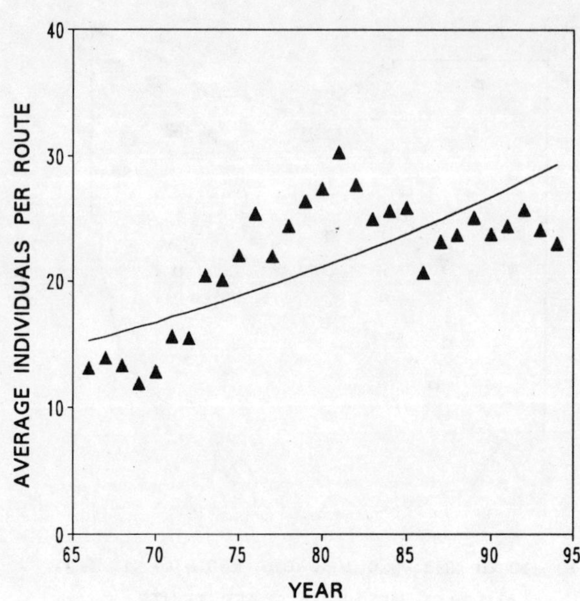

MIGRATORY STATUS: Central neotropical migrant.

BREEDING HABITAT: Scrub, including brushy abandoned fields, corridors, and edges.

ABUNDANCE AND DISTRIBUTION: Abundant (21.5 birds per route) and widely distributed (all routes). More abundant in Eastern than Western Ohio (26.5 vs. 17.2 birds per route, P = 0.03) and in Southern than Northern Ohio (27.6 vs. 12.5 birds per route, P = 0.0001).

OHIO POPULATION TREND: Percent Annual Change (± SE) = 2.4 (± 0.5), P < 0.001

Indigo Buntings increased significantly at 2.4% annually. The increase was pronounced through 1981 (6.5% annually, P < 0.001) but numbers from 1982-1994 have been fairly stable (-0.6%, P = 0.28).

The overall rate of increase was similar in Western and Eastern Ohio (2.6 vs. 2.2%, P = 0.64) but was significantly higher in Northern than Southern Ohio (4.0 vs. 1.7%, P = 0.04).

CONTINENTAL AND GREAT LAKES REGIONAL TREND: In contrast, the Great Lakes population did not exhibit a significant trend (-0.2% annual change), and the continental population decreased significantly at -0.6% annually.

DICKCISSEL

Spiza americana

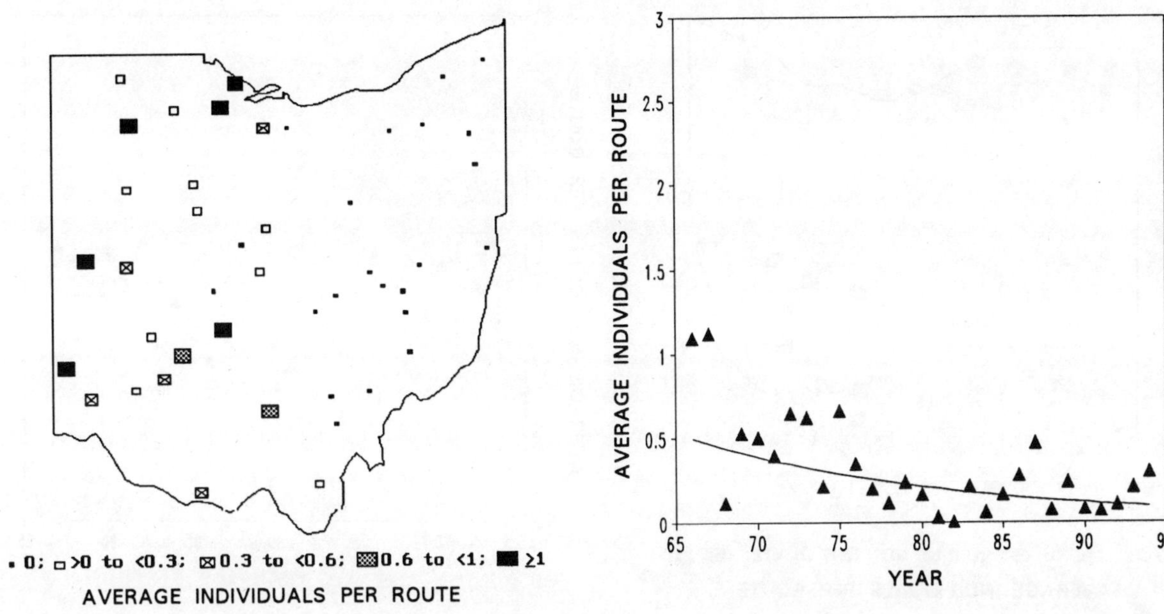

MIGRATORY STATUS: Southern neotropical migrant.

BREEDING HABITAT: Grasslands, particularly hayfields of alfalfa or clover, fallow fields, and pastures.

ABUNDANCE AND DISTRIBUTION: Rare (0.3 birds per route) and locally distributed (23 routes). Much more common in Western than Eastern Ohio (0.56 vs. 0.05 birds per route, P < 0.001); recorded on only 2 routes in Eastern Ohio.

OHIO POPULATION TREND: Percent Annual Change (± SE) = -5.9 (± 1.7), P = 0.002
Dickcissels declined significantly at 5.9% annually. If the particularly high initial counts in 1966 and 1967 are excluded, the decline of 4.6% annually remains significant (P = 0.02). The large variation about the regression line is indicative of the temporal and spatial variation in density typical of Dickcissels. Too rare in Eastern Ohio to allow a comparison of Eastern vs. Western trends.

CONTINENTAL AND GREAT LAKES REGIONAL TREND: Similarly, Dickcissels in the regional and continental populations have declined significantly since 1966 at 3.6 and 1.6% annually.

RUFOUS-SIDED TOWHEE

Pipilo erythrophthalmus

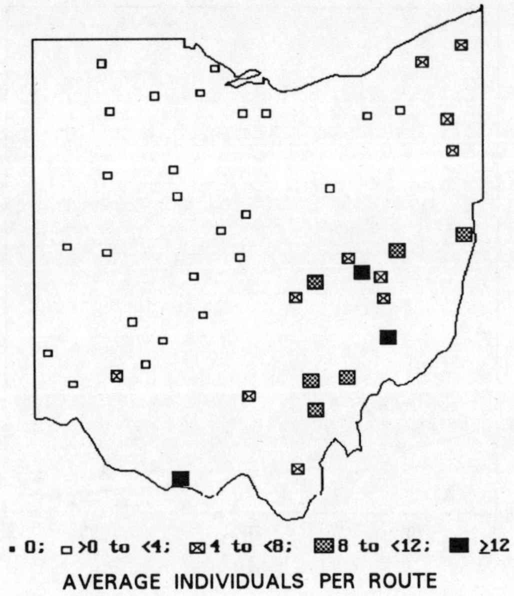

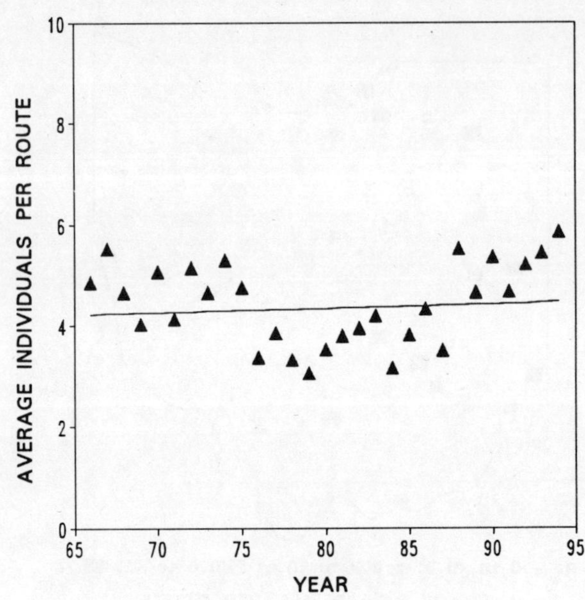

MIGRATORY STATUS: Temperate migrant, although some winter in southern Ohio.

BREEDING HABITAT: Scrub, including brushy successional habitat, edges, and woodland openings.

ABUNDANCE AND DISTRIBUTION: Common (4.4 birds per route) and widely distributed (45 routes).
Much more common in Eastern than Western Ohio (7.8 vs. 1.4 birds per route, P < 0.001).

OHIO POPULATION TREND: **Percent Annual Change (± SE) = 0.2 (± 0.6), P = 0.75**
   Ohio's Rufous-sided Towhee population did not exhibit a significant overall trend (0.2% annual change).
Numbers were stable through 1975 (0.0%, P < 0.001), declined sharply, and then increased significantly at
3.1% annually during 1976-1994 (P < 0.001). Note that the severe winters of 1976-1978 may have
contributed to the low numbers during the late 1970s, but that the decline appears to have begun the previous
spring (1976).
   Eastern and Western trends were not significantly different (0.03 vs. 0.8%, P = 0.37).

CONTINENTAL AND GREAT LAKES REGIONAL TREND: Rufous-sided Towhees declined
significantly in both the Great Lakes Region and continent-wide at 1.6 and 1.7% annually.

CHIPPING SPARROW

Spizella passerina

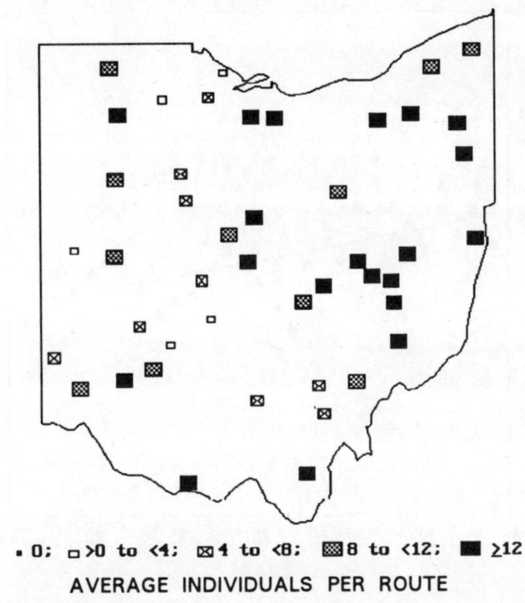

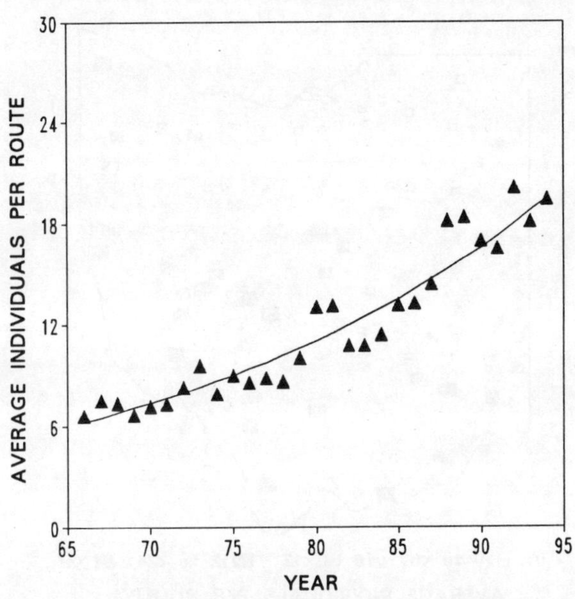

MIGRATORY STATUS: Temperate migrant; some winter in Mexico and the Caribbean.

BREEDING HABITAT: Residential areas, especially near large grassy areas.

ABUNDANCE AND DISTRIBUTION: Common (11.8 birds per route) and widely distributed (all routes). Nearly twice as common in Eastern than Western Ohio (15.4 vs. 8.7 birds per route, P < 0.001).

OHIO POPULATION TREND: Percent Annual Change (± SE) = 4.1 (± 0.8), P < 0.001
   Chipping Sparrows have increased dramatically and significantly at 4.1% annually. The increase probably results from an increase in urban habitat. Chipping Sparrows have increased significantly in both Western (7.2%, P < 0.001) and Eastern Ohio (2.4%, P = 0.02), although the Western increase was significantly greater (P = 0.006).

CONTINENTAL AND GREAT LAKES REGIONAL TREND: Chipping Sparrows also increased significantly in the Great Lakes Region at 1.9% annually. However, the continental population did not exhibit a significant overall trend (-0.2% annual change).

FIELD SPARROW

Spizella pusilla

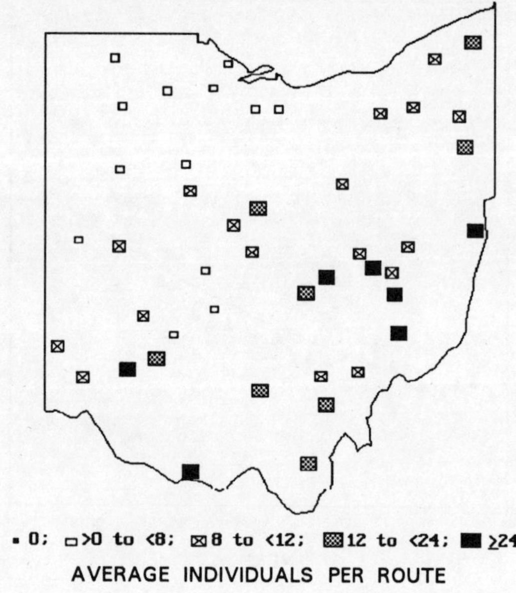

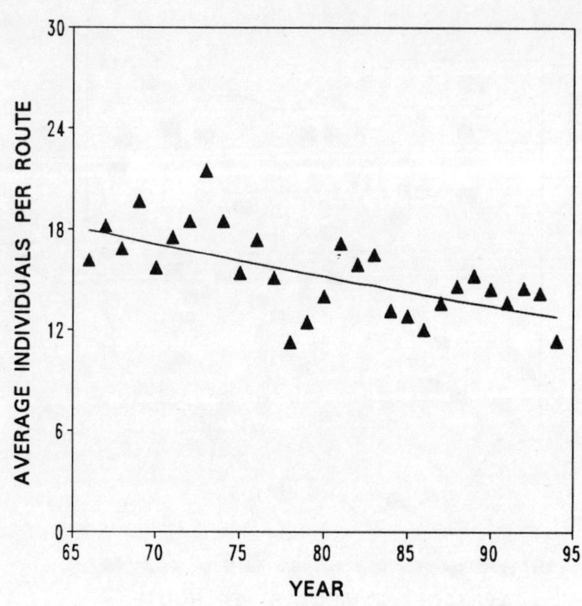

MIGRATORY STATUS: Temperate migrant.

BREEDING HABITAT: Weedy grasslands, particularly where herbaceous and woody vegetation is interspersed.

ABUNDANCE AND DISTRIBUTION: Abundant (15.3 birds per route) and widely distributed (all routes). More common in Eastern than Western Ohio (19.5 vs. 11.7 birds per route, $P = 0.01$).

OHIO POPULATION TREND: Percent Annual Change ($\pm$ SE) = -1.2 ($\pm$ 0.5), $P = 0.02$

Field Sparrows have declined significantly at 1.2% annually. Declines in Eastern and Western Ohio did not differ significantly (1.4 vs. 0.9%, $P = 0.59$).

The severe winters of 1976-78 reduced numbers noticeably, but a decline was evident both before and after those winters. The decline probably results from habitat loss to more intensive agriculture in the west and to maturation of forests in the east.

CONTINENTAL AND GREAT LAKES REGIONAL TREND: Similarly, Field Sparrows in the Great Lakes Region and continent-wide have declined rapidly and significantly at 3.0 and 3.4% annually. The decline is thought to be due, in part, to more intensive land use and a decreased availability of reverting fields (Robbins et al., 1986).

VESPER SPARROW

Pooecetes gramineus

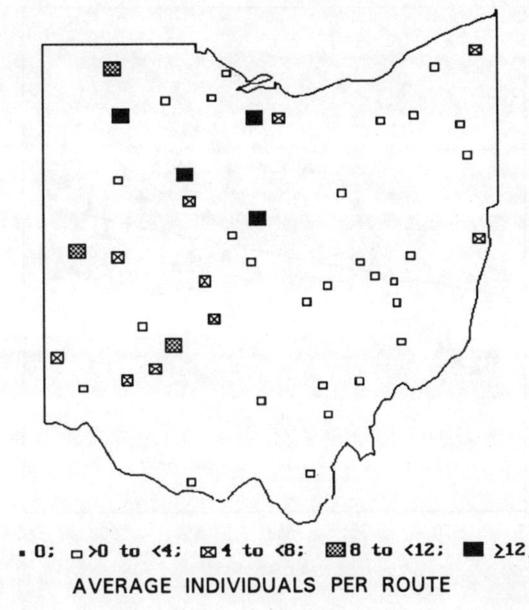

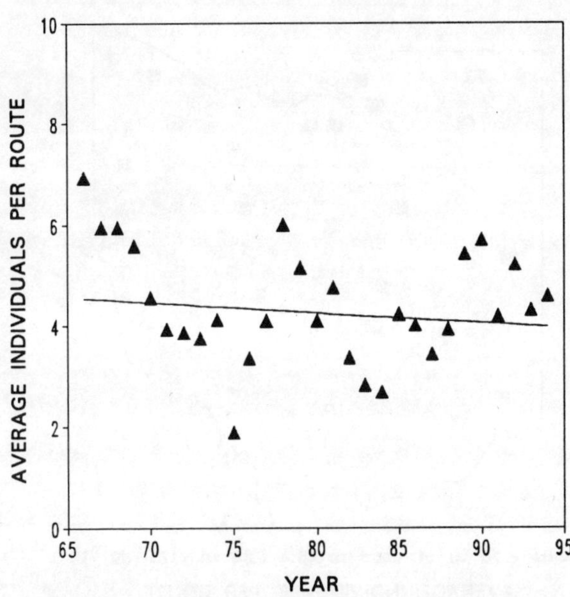

MIGRATORY STATUS: Temperate migrant.

BREEDING HABITAT: Grasslands, especially short-grass upland meadows.

ABUNDANCE AND DISTRIBUTION: Common (4.4 birds per route) and widely distributed (all routes). Much more common in the grasslands of Western Ohio than in Eastern Ohio (7.1 vs. 1.4 birds per route, $P <$ 0.001).

OHIO POPULATION TREND: Percent Annual Change ($\pm$ SE) = **-0.5 ($\pm$ 1.8), P = 0.80**

Vesper Sparrows have not exhibited a significant overall annual trend (-0.5%). Note that the cyclic tendency and high annual variation exhibited by Vesper Sparrows would make any existing trend difficult to detect. Vesper Sparrows decreased significantly in Eastern Ohio at 6.2% annually (P = 0.04) but showed high variation and no significant trend in Western Ohio (0.4%, P = 0.85)

CONTINENTAL AND GREAT LAKES REGIONAL TREND: The regional and continental populations have declined significantly at 1.7 and 0.6% annually.

SAVANNAH SPARROW

Passerculus sandwichensis

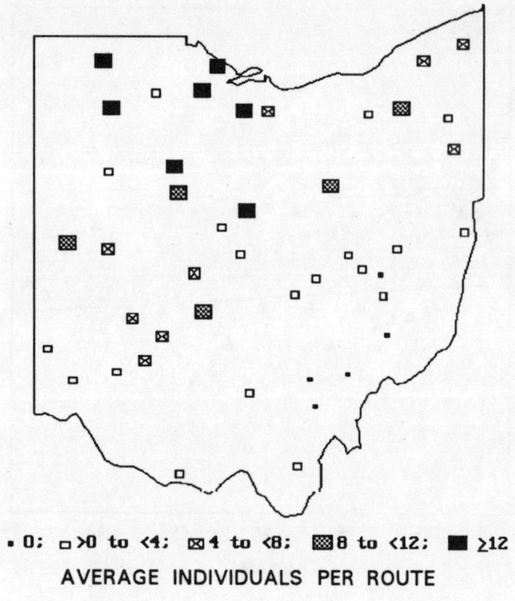

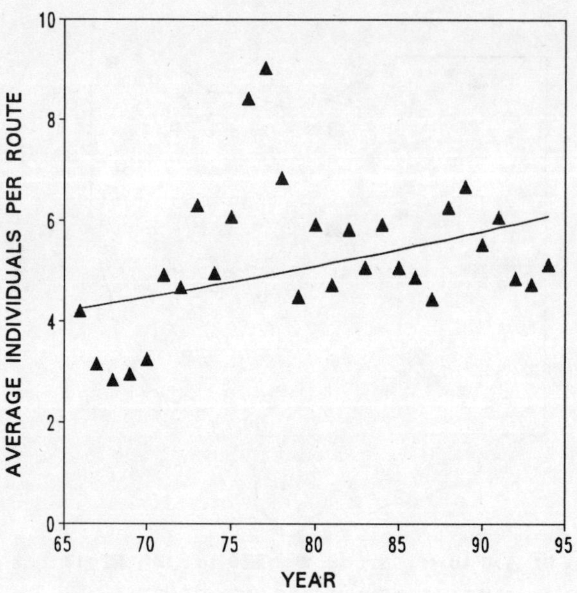

MIGRATORY STATUS: Temperate migrant.

BREEDING HABITAT: Grasslands.

ABUNDANCE AND DISTRIBUTION: Common (5.3 birds per route) and widely distributed (40 routes). Nearly 4 times more common in the grasslands of Western Ohio than in Eastern Ohio (7.8 vs. 2.3, P = 0.001). Savannah Sparrows had spread southward into Northern Ohio by the 1920s, into southwestern Ohio by the late 1940s, into southeastern Ohio by the 1970s (Peterjohn, 1989), and are still more common in Northern than Southern Ohio (8.5 vs. 2.4 birds per route, P = 0.001).

OHIO POPULATION TREND: Percent Annual Change (± SE) = 1.3 (± 1.2), P = 0.27
Savannah Sparrows have increased at 1.3% annually, but the overall trend is not significant because of large annual variation. The population is best described as having increased significantly at 9.9% annually (P = 0.03) during 1966-1977 and as having remained fairly stable during 1978-1994 (-0.3% annual change, P = 0.81).

Trends for Western and Eastern Ohio did not differ significantly (0.5 vs. 4.1%, P = 0.72). The increase in the Savannah Sparrow population, despite a loss of grassland habitat, may be influenced by their range expansion into Ohio.

CONTINENTAL AND GREAT LAKES REGIONAL TREND: Both the regional and continental populations declined significantly at 1.1 and -0.6% annually.

GRASSHOPPER SPARROW

Ammodramus savannarum

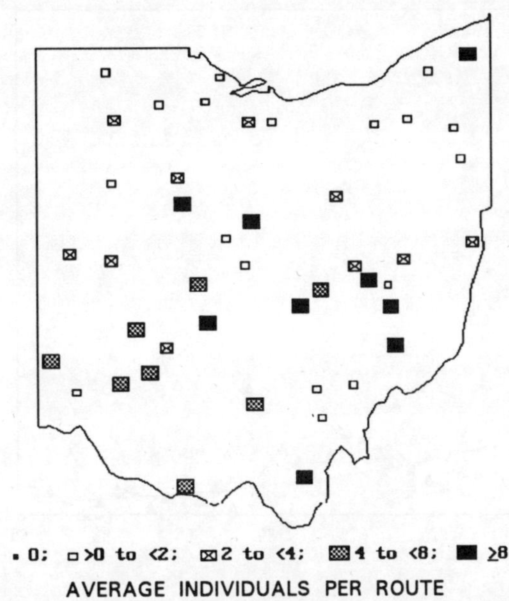

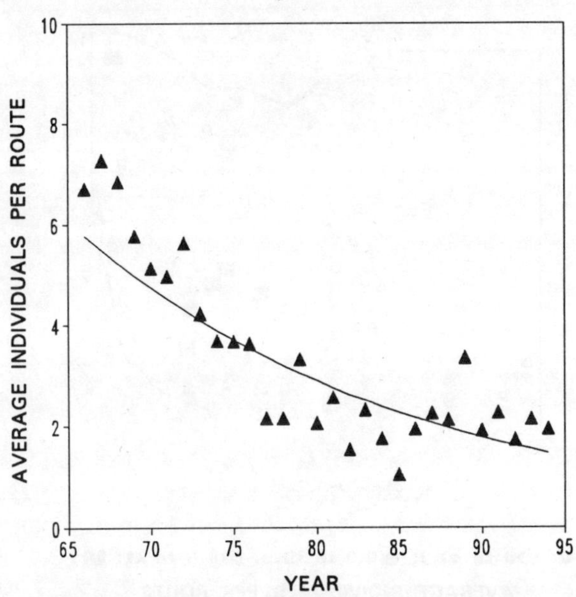

MIGRATORY STATUS: Central neotropical migrant.

BREEDING HABITAT: Grasslands, especially extensive tall-grass pastures and hayfields.

ABUNDANCE AND DISTRIBUTION: Fairly common (3.4 birds per route) and widely distributed (all routes). Equally abundant in Western and Eastern Ohio (3.2 vs. 3.7 birds per route, P = 0.65).

OHIO POPULATION TREND: **Percent Annual Change ($\pm$ SE) = -4.8 ($\pm$ 1.0), P < 0.001**
   Grasshopper Sparrows have declined dramatically and significantly at 4.8% annually. Declines in Western and Eastern Ohio were not significantly different (-3.4 vs. -6.6%, P = 0.10). Like most other grassland birds in Ohio, Grasshopper Sparrows are probably declining due to habitat loss.

CONTINENTAL AND GREAT LAKES REGIONAL TREND: The regional and continental populations have also declined dramatically and significantly at 5.5 and 3.6% annually.

HENSLOW'S SPARROW

Ammodramus henslowii

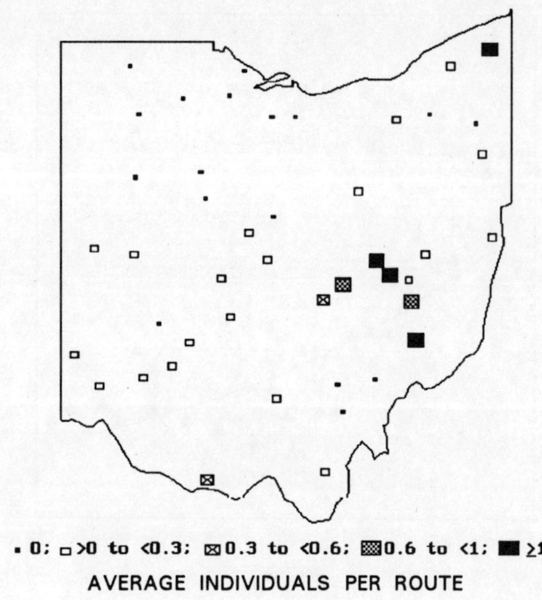

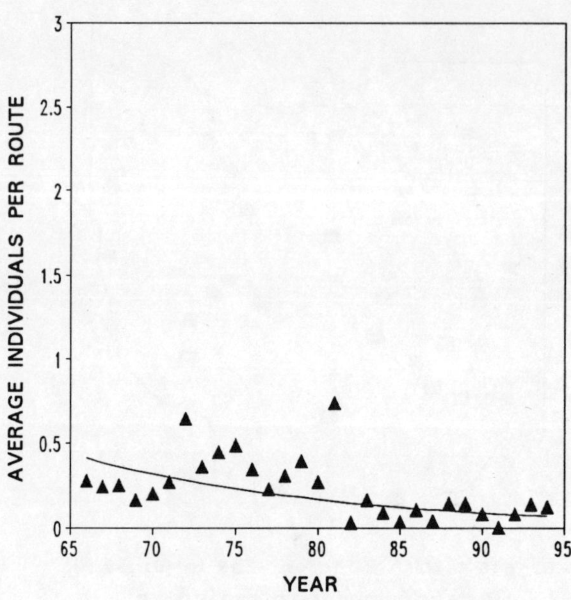

MIGRATORY STATUS: Temperate migrant.

BREEDING HABITAT: Grasslands, including extensive fields of predominantly grasses and early successional habitats.

ABUNDANCE AND DISTRIBUTION: Rare (0.2 birds per route) and locally distributed (28 routes). More common in Eastern than Western Ohio (0.465 vs. 0.05 birds per route, P = 0.0078).

OHIO POPULATION TREND: Percent Annual Change (± SE) = -6.2 (± 3.0), P = 0.05
   Henslow's Sparrows declined at a substantial rate of 6.2% annually, but the trend was only marginally statistically significant. Note that the high annual variation would make any existing trend difficult to detect. Declines in Eastern and Western Ohio were similar (-6.2 vs. -6.0%, P = 0.96). The loss of grassland habitat is probably the primary cause of the decline.

CONTINENTAL AND GREAT LAKES REGIONAL TREND: The regional and continental populations have also experienced large and significant declines of 7.6 and 8.3% annually.

SONG SPARROW

Melospiza melodia

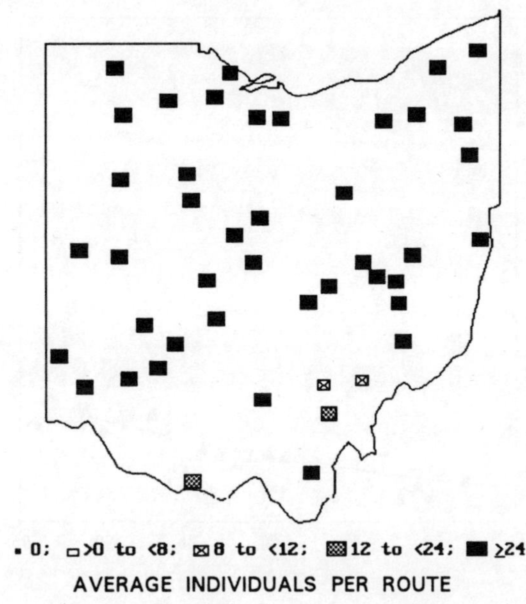

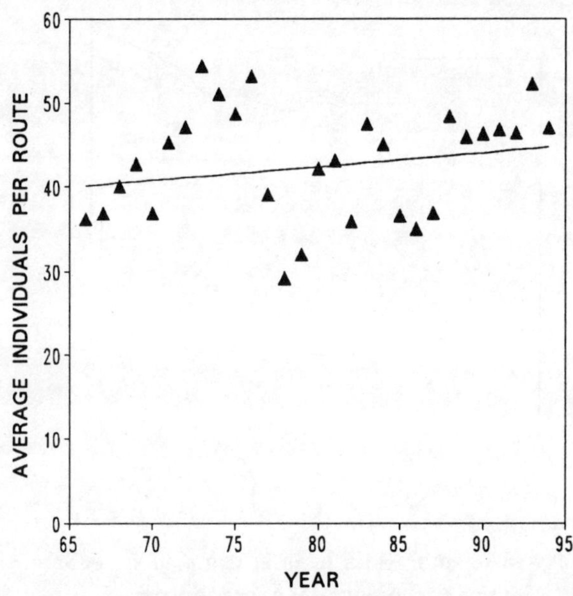

MIGRATORY STATUS: Temperate migrant.

BREEDING HABITAT: Shrub, including thickets bordering woodlands, grasslands, and residential areas.

ABUNDANCE AND DISTRIBUTION: Abundant (43.4 birds per route) and widely distributed (all routes). Somewhat, but not significantly, more abundant in Western than Eastern Ohio (48.0 vs. 38.2, P = 0.07).

OHIO POPULATION TREND: **Percent Annual Change (± SE) = 0.4 (± 0.6), P = 0.50**
   The overall 1966-1994 population trend is not significant (0.4% annual change). Song Sparrows were increasing significantly at 4.2% annually prior to the severe winters of 1976-78 (1966-1976, P = 0.003) and have increased significantly at 2.0% annually thereafter (1977-1994, P = 0.007), although annual counts have fluctuated greatly.
   Overall trends for Eastern and Western Ohio did not differ significantly (1.4 vs. -0.3%, P = 0.16).

CONTINENTAL AND GREAT LAKES REGIONAL TREND: The regional population exhibited a small but significant increase of 0.5% annually and the continental population exhibited a small but significant decrease of 0.7% annually.

SWAMP SPARROW

Melospiza georgiana

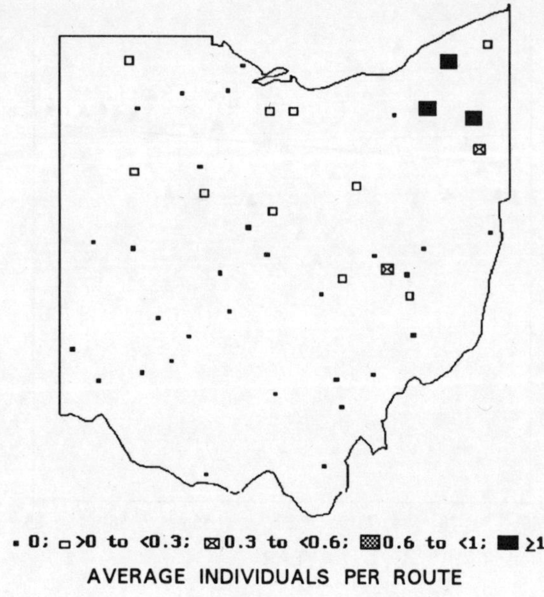

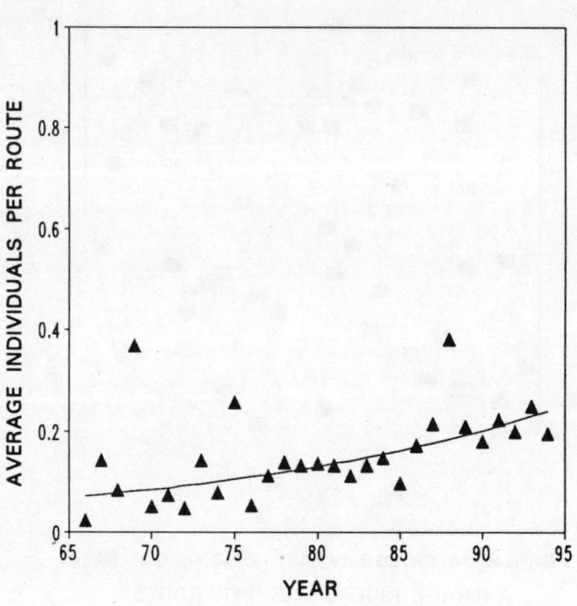

MIGRATORY STATUS: Temperate migrant.

BREEDING HABITAT: Wetlands, including brushy or grassy margins of marshes, ponds, and lakes.

ABUNDANCE AND DISTRIBUTION: Rare (0.15 birds per route) and very locally distributed (15 routes). More common in northeastern Ohio in the Glaciated Plateau than elsewhere (0.77 vs. 0.03 birds per route, $P = 0.05$); the southern third of Ohio is outside of their typical breeding range. Thought to have expanded their range slightly southward between 1940-80 (Peterjohn and Rice, 1991).

OHIO POPULATION TREND: Percent Annual Change ($\pm$ SE) = 4.4 ($\pm$ 2.8), P = 0.12
   Swamp Sparrows appear to have increased substantially (4.4% annually), but the trend is not statistically significant (note particularly high counts in 1969, 1975, and 1988). Consistent with their southward range expansion, Swamp Sparrows increased significantly in Southern Ohio (10.0%, $P = 0.01$) but not Northern Ohio (3.8%, $P = 0.25$); both trends exhibited substantial annual variation.

CONTINENTAL AND GREAT LAKES REGIONAL TREND: The regional population increased significantly at 2.6% annually, but the continental population did not exhibit a significant overall trend (0.4% annual change).

BOBOLINK

Dolichonyx oryzivorus

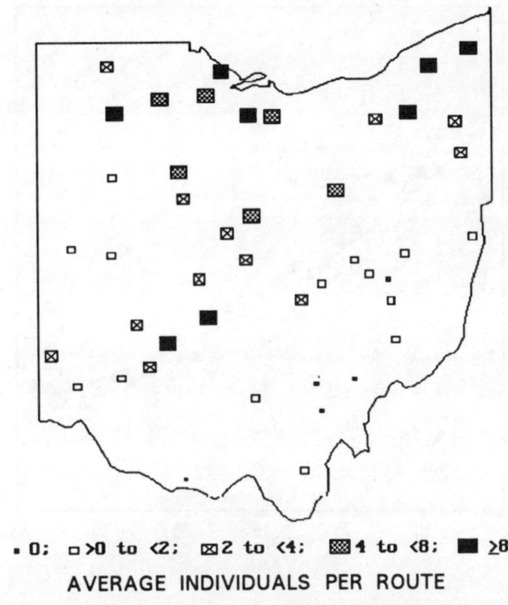

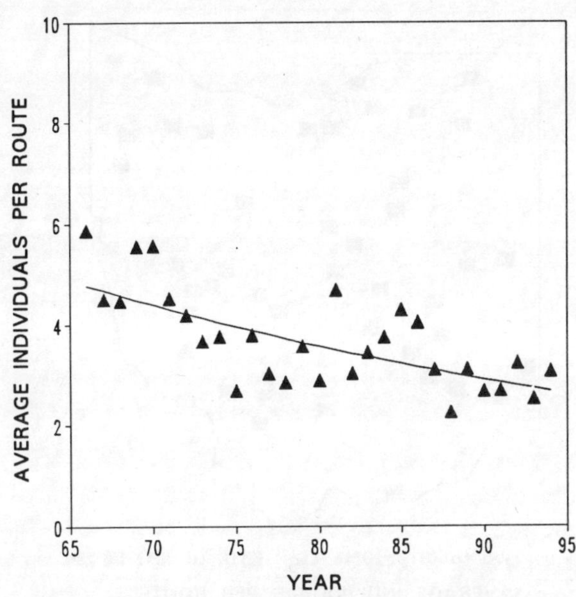

MIGRATORY STATUS: Southern neotropical migrant.

BREEDING HABITAT: Grasslands, especially hayfields and early successional fields of grasses and weeds.

ABUNDANCE AND DISTRIBUTION: Fairly common (3.5 birds per route) and widely distributed (40 routes). Ohio is on the southern edge of their breeding range, and Bobolinks were more common in Northern than Southern Ohio (5.7 vs. 1.6 birds per route, P < 0.001); especially rare in the southeastern quarter of Ohio (Unglaciated Plateau, 0.3 birds per route).

OHIO POPULATION TREND: Percent Annual Change ($\pm$ SE) = -2.0 ($\pm$ 1.0), P = 0.04
  Bobolinks have declined significantly at 2.0% annually. Bobolinks declined significantly in Western -3.0%, P = 0.04) but not Eastern Ohio (-0.5%, P = 0.67); and they declined significantly in Southern (-6.6%, P < 0.001) but not Northern Ohio (-0.4%, P = 0.62) despite being more common in Northern Ohio.
  The decline of Bobolinks in Ohio and elsewhere is attributed to conversion of grasslands to row crops and more frequent mowing of hayfields (Peterjohn, 1989).

CONTINENTAL AND GREAT LAKES REGIONAL TREND: The regional and continental populations have also declined significantly at 3.3 and 1.6% annually.

RED-WINGED BLACKBIRD

Agelaius phoeniceus

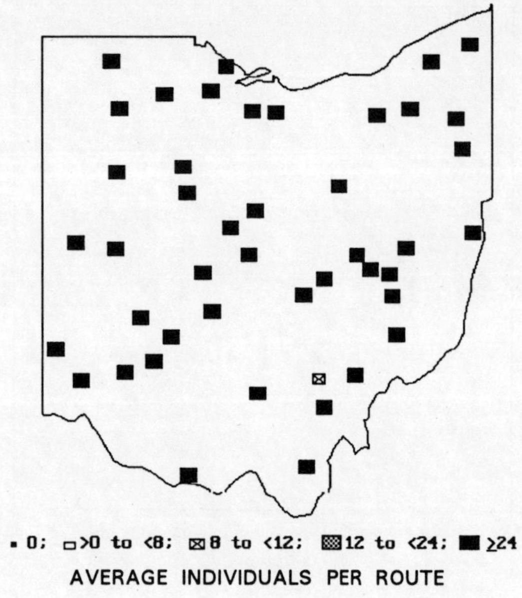

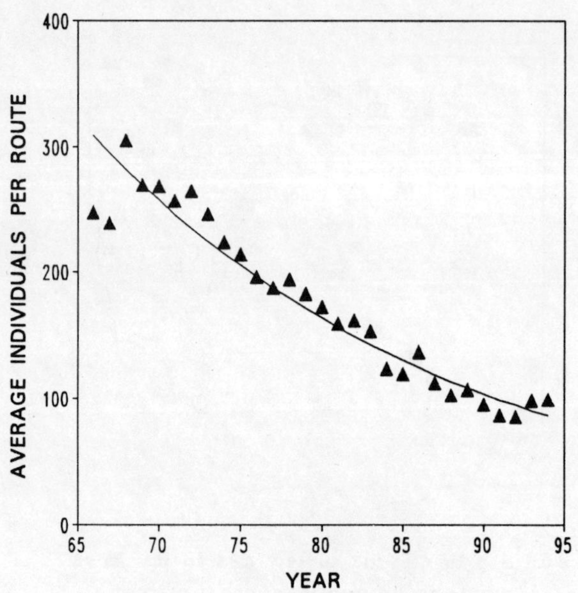

MIGRATORY STATUS: Temperate migrant.

BREEDING HABITAT: Wetlands, particularly cattail marshes. Use of upland grassy pastures and hayfields has become more common as wetlands are drained.

ABUNDANCE AND DISTRIBUTION: Very abundant (177 birds per route) and widely distributed (all routes). Abundance in Western and Eastern Ohio not significantly different (204 vs. 147 birds per route, P = 0.11).

OHIO POPULATION TREND: Percent Annual Change (± SE) = -4.5 (± 0.6), P < 0.001
   Red-winged Blackbirds have decreased dramatically and significantly at 4.5% annually. Red-winged Blackbirds are thought to have increased during much of the 1950s and 1960s in eastern North America, becoming targets of control efforts in the 1970s. The loss of wetland and grassland habitat throughout the 1900s is probably responsible for the decline of Red-winged Blackbirds.
   Trends in Western and Eastern Ohio did not differ significantly (-4.9 vs. -3.8%, P = 0.52).

CONTINENTAL AND GREAT LAKES REGIONAL TREND: Similarly, the regional and continental populations have decreased significantly at 1.2 and 1.1% annually.

EASTERN MEADOWLARK

Sturnella magna

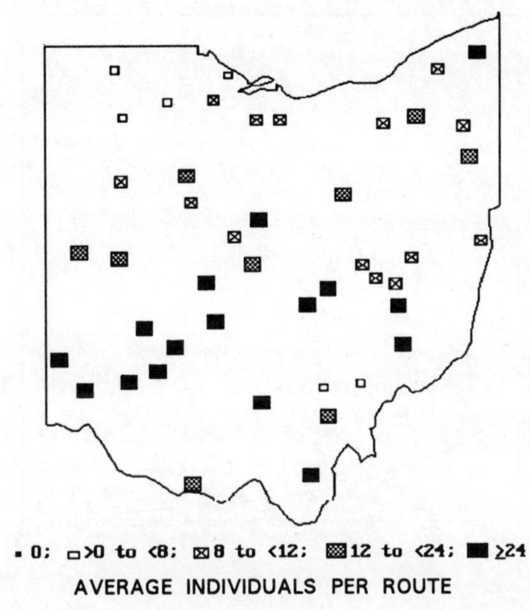

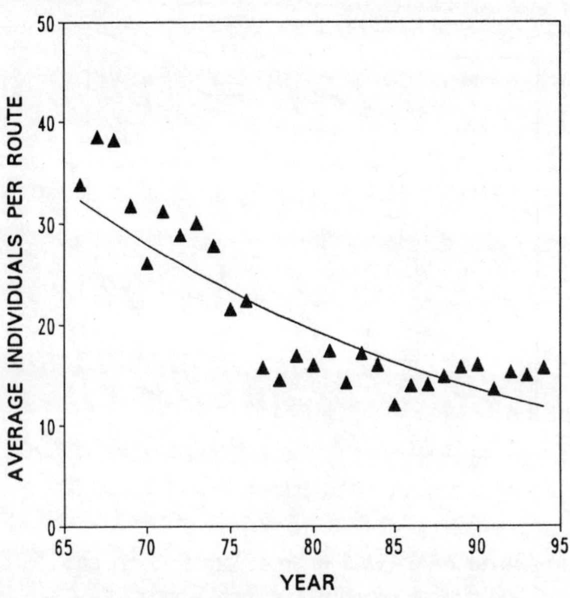

MIGRATORY STATUS: Temperate migrant.

BREEDING HABITAT: Grasslands, including pastures, hayfields, and weedy fallow fields.

ABUNDANCE AND DISTRIBUTION: Abundant (21.1 birds per route) and widely distributed (all routes). Abundance in Western and Eastern Ohio did not differ significantly (21.8 vs. 20.3 birds per route, P = 0.71).

OHIO POPULATION TREND: Percent Annual Change ($\pm$ SE) = -3.5 ($\pm$ 0.4), P < 0.001

Eastern Meadowlarks have decreased dramatically and significantly at 3.5% annually. The decline is thought to have begun in the 1940s and be due to habitat loss (Peterjohn, 1989).

Trends in Eastern and Western Ohio were similar (-3.2 vs. -3.8%, P = 0.47).

CONTINENTAL AND GREAT LAKES REGIONAL TREND: Similarly, the regional and continental populations have experienced large and significant declines of 2.9 and 2.6% annually.

WESTERN MEADOWLARK

Sturnella neglecta

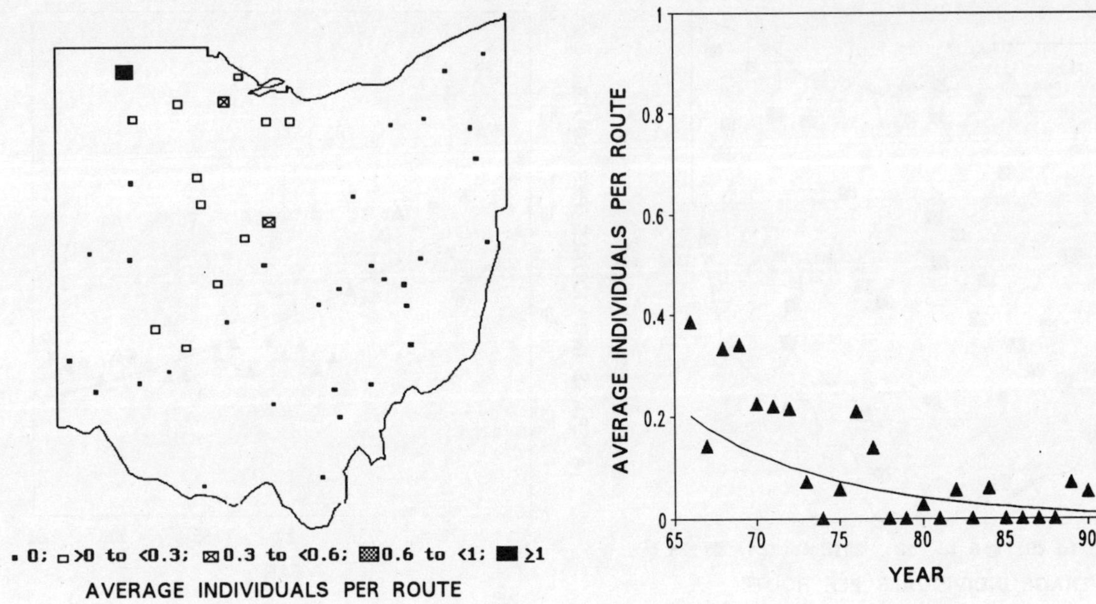

MIGRATORY STATUS: Temperate migrant.

BREEDING HABITAT: Grasslands, including pastures, hayfields, and weedy fallow fields.

ABUNDANCE AND DISTRIBUTION: Rare (0.1 birds per route) and very locally distributed (14 routes). Western Meadowlarks occur only on routes in Western Ohio.

OHIO POPULATION TREND: Percent Annual Change (± SE) = -10.8 (± 2.6), P = 0.001

Western Meadowlarks have declined significantly at 10.8% annually; they appear to have declined sharply between 1966-1978 and to have remained very rare since then. Because Western Meadowlarks are recorded on few routes and in low abundance, the exact value of the trend should be interpreted with caution.

CONTINENTAL AND GREAT LAKES REGIONAL TREND: Similarly, the regional and continental populations have declined significantly at 4.0 and 0.7% annually.

COMMON GRACKLE

Quiscalus quiscula

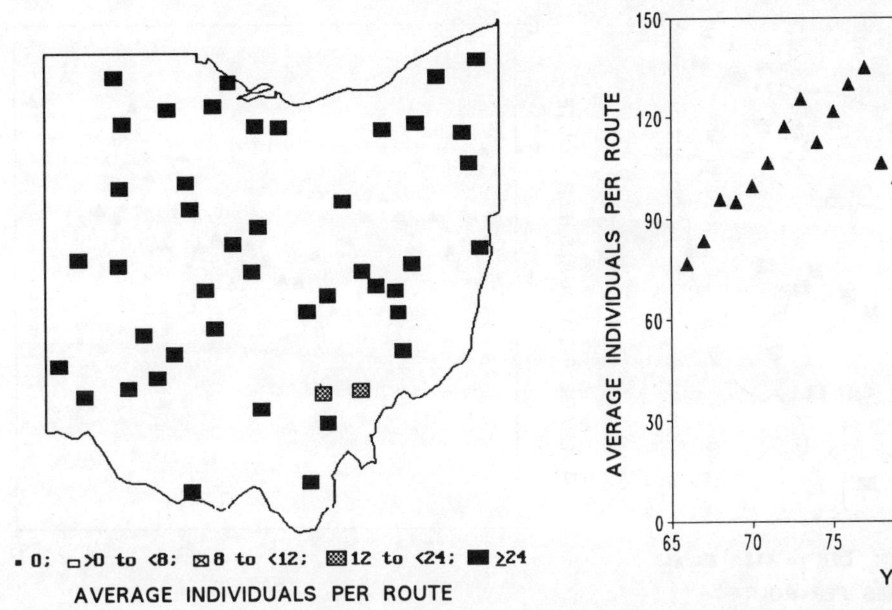

MIGRATORY STATUS: Temperate migrant.

BREEDING HABITAT: Shrub, including shrubby edges along pastures, woodlands, and residential areas.

ABUNDANCE AND DISTRIBUTION: Very abundant (97.8 birds per route) and widely distributed (all routes). More abundant in Western than Eastern Ohio (119 vs. 74 birds per route, P = 0.008).

OHIO POPULATION TREND: PERCENT ANNUAL CHANGE calculated 1966-1994 not appropriate.
   Common Grackles increased sharply at 4.6% annually through 1977 (P = 0.001), and declined sharply thereafter at 2.6% annually (1978-1994, P = 0.002). Common Grackles decreased significantly in Western Ohio at 1.4% annually (P = 0.03) but showed large annual variation and no significant trend in Eastern Ohio (0.1%, P = 0.85).
   Like other blackbirds, Common Grackles increased during the 1940s and 1950s in eastern North American (Peterjohn, 1989).

CONTINENTAL AND GREAT LAKES REGIONAL TREND: The regional and continental populations decreased significantly at 1.3 and 1.5% annually. Note that the Common Grackle's regional and continental trends may be misleading if the strongly non-exponential pattern observed in Ohio is evident elsewhere.

BROWN-HEADED COWBIRD

Molothrus ater

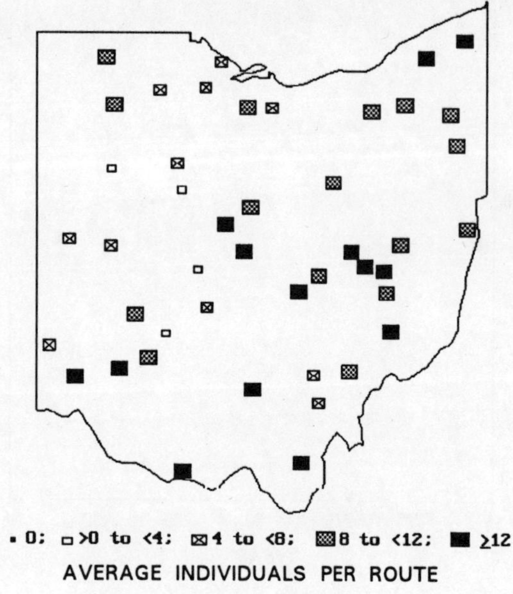

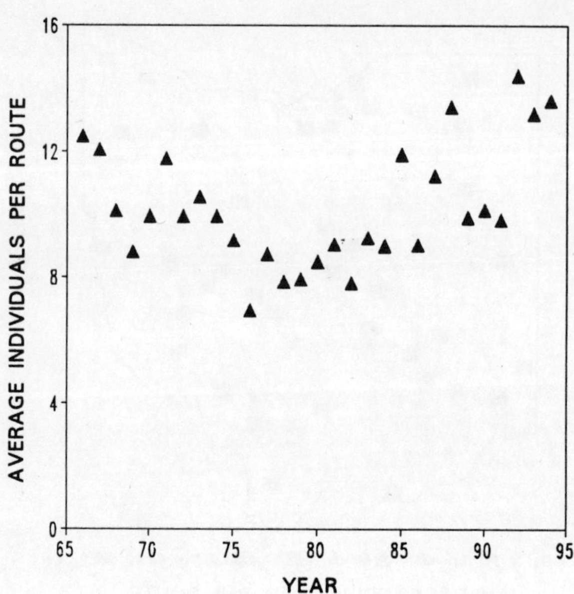

MIGRATORY STATUS: Temperate migrant.

BREEDING HABITAT: Edges of woodlands and agricultural fields, fallow fields, and residential areas.

ABUNDANCE AND DISTRIBUTION: Common (10.2 birds per route) and widely distributed (all routes). More common in Eastern than Western Ohio (12.4 vs. 8.3 birds per route, P = 0.004).

Cowbirds expanded into Ohio in the 1840s, had become widely distributed by the late 1800s, and increased substantially during the 1950s and 60s.

OHIO POPULATION TREND: Percent Annual Change calculated for **1966-1994 not appropriate.**

Brown-headed Cowbirds declined at 3.1% annually through 1979 (P = 0.001) and increased at 3.4% annually thereafter (1980-1994, P = 0.003). In Western Ohio, Brown-headed Cowbirds have increased somewhat, but not significantly since 1966, at 1.6% annually (P = 0.06), but exhibited large annual variation and no significant trend in Eastern Ohio (-0.2%, P = 0.86).

CONTINENTAL AND GREAT LAKES REGIONAL TREND: The regional and continental populations decreased modestly, but significantly, at 0.6 and 0.9% annually. Note that the Brown-headed Cowbird's regional and continental trends may be misleading if the strongly non-exponential pattern observed in Ohio is evident elsewhere.

ORCHARD ORIOLE

Icterus spurius

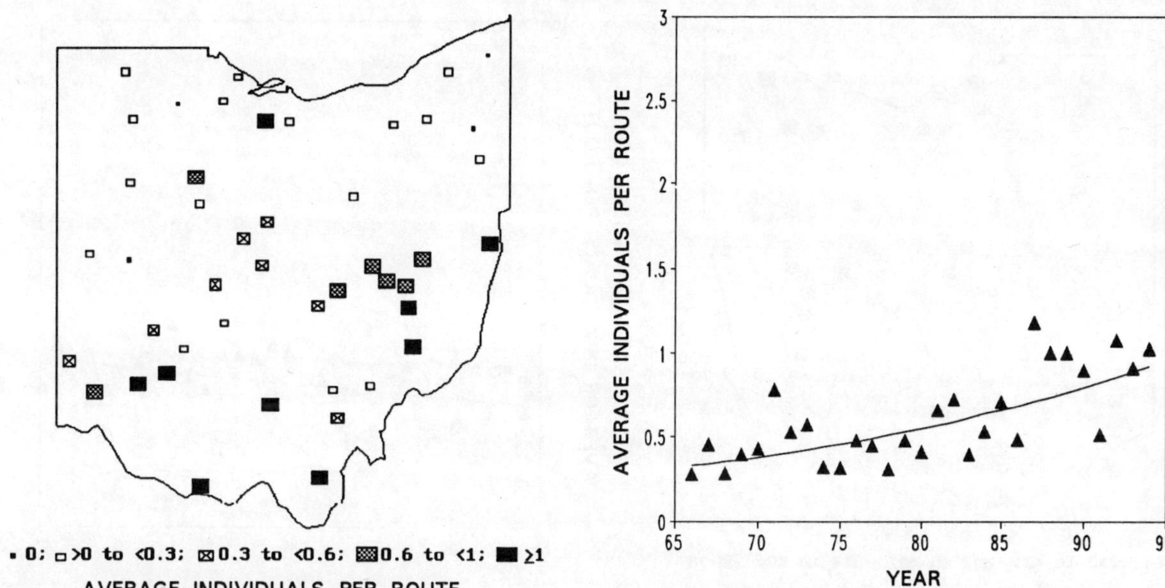

MIGRATORY STATUS: Central neotropical migrant.

BREEDING HABITAT: Young riparian corridors or woodland edges with little undergrowth and grassy park-like areas interspersed with young trees.

ABUNDANCE AND DISTRIBUTION: Rare (0.6 birds per route) and widely distributed (41 routes). More common in Southern than Northern Ohio (0.8 vs. 0.3 birds per route, $P = 0.006$). Equally rare in Western and Eastern Ohio (0.5 vs. 0.7, $P = 0.47$).

OHIO POPULATION TREND: Percent Annual Change ($\pm$ SE) = 3.8 ($\pm$ 1.5), **P = 0.02**
  Orchard Orioles have increased significantly at 3.8% annually. Trends were similar in Western vs. Eastern (3.7 vs. 4.0%, $P = 0.92$). Orchard Orioles increased dramatically and significantly in Northern Ohio (10.1%, $P < 0.001$) but did not exhibit a significant trend in Southern Ohio (2.2%, $P = 0.20$).

CONTINENTAL AND GREAT LAKES REGIONAL TREND: In contrast, the continental population decreased significantly at 1.9% annually and the Great Lakes population did not exhibit a significant trend.

NORTHERN ORIOLE

Icterus galbula

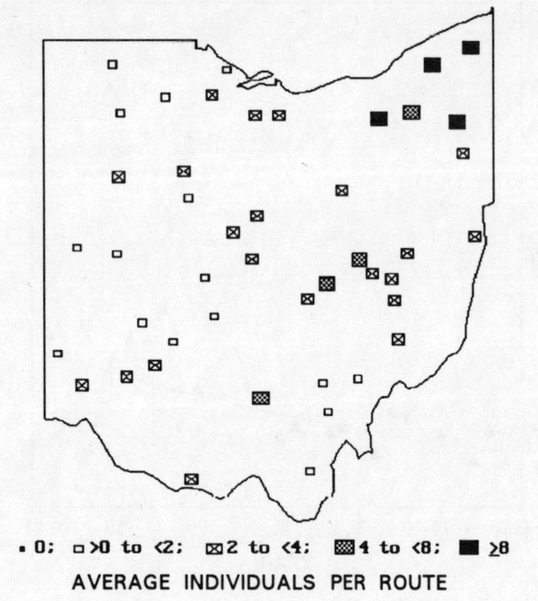

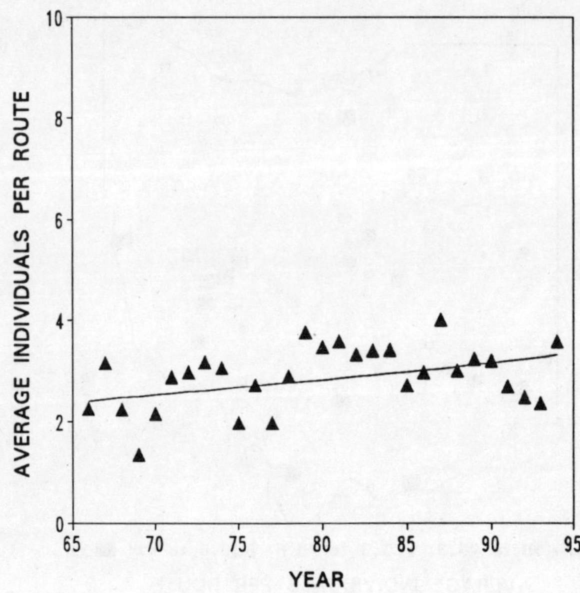

MIGRATORY STATUS: Central neotropical migrant.

BREEDING HABITAT: Mature riparian corridors or woodland edges with little undergrowth, and grassy park-like areas interspersed with mature trees.

ABUNDANCE AND DISTRIBUTION: Fairly common (2.8 birds per route) and widely distributed (all routes). More common in Eastern than Western Ohio (3.7 vs. 2.0 birds per route, P = 0.003).

OHIO POPULATION TREND: Percent Annual Change (± SE) = 1.1 (± 0.9), P = 0.20
Northern Orioles did not exhibit a significant annual increase (1.1%). Northern Orioles increased significantly in Western Ohio (3.5%, P < 0.001) but not in Eastern Ohio (-0.2%, P = 0.87).

CONTINENTAL AND GREAT LAKES REGIONAL TREND: The eastern subspecies of the Northern Oriole (Baltimore Oriole) did not exhibit a significant trend in either the Great Lakes or continental population.

HOUSE FINCH

Carpodacus mexicanus

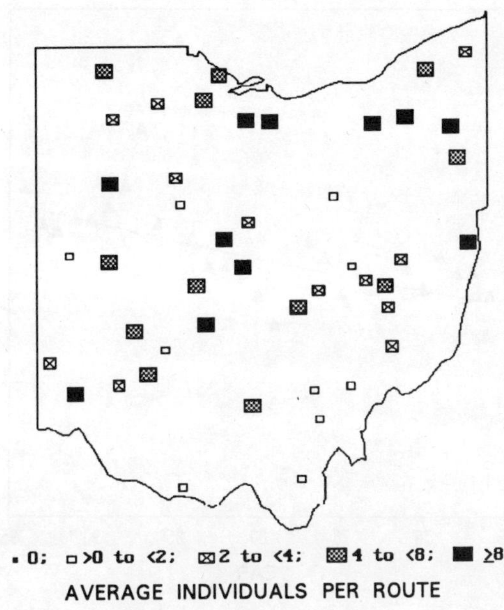

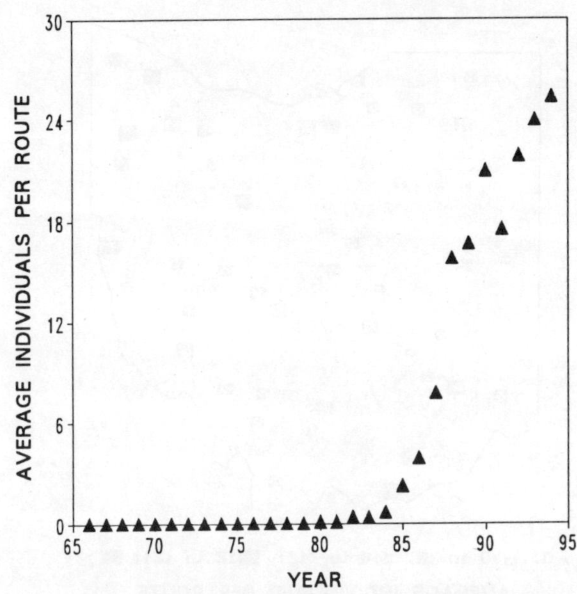

MIGRATORY STATUS: Permanent resident.

BREEDING HABITAT: Residential areas, including rural and urban areas.

ABUNDANCE AND DISTRIBUTION: In Ohio, House Finches were thought to have begun nesting in the late 1970s and were first recorded on a BBS route in 1980. Today House Finches are abundant (25 birds per route in 1994) and widely distributed (all routes).

OHIO POPULATION TREND: **Percent Annual Change (± SE) = 57.3 (± 1.7), P < 0.001**
  House Finches have increased dramatically at 57.3% annually since their first occurrence on a BBS route (1980-1991, P < 0.001). Trends in Eastern and Western Ohio were similar (57.5 vs. 56.2%, P = 0.83).

CONTINENTAL AND GREAT LAKES REGIONAL TREND: The Eastern U.S. population, which spread from an introduction in New York City, increased dramatically during 1966-1979 at 58% annually and has grown somewhat more slowly during the last 15 years (1980-1994, 18.9%). Similarly the Great Lakes population has increased dramatically at 48.9% since House Finches arrived in the region in 1980. The continental population has also increased significantly, although less dramatically at 1.7% annually.

AMERICAN GOLDFINCH

Carduelis tristis

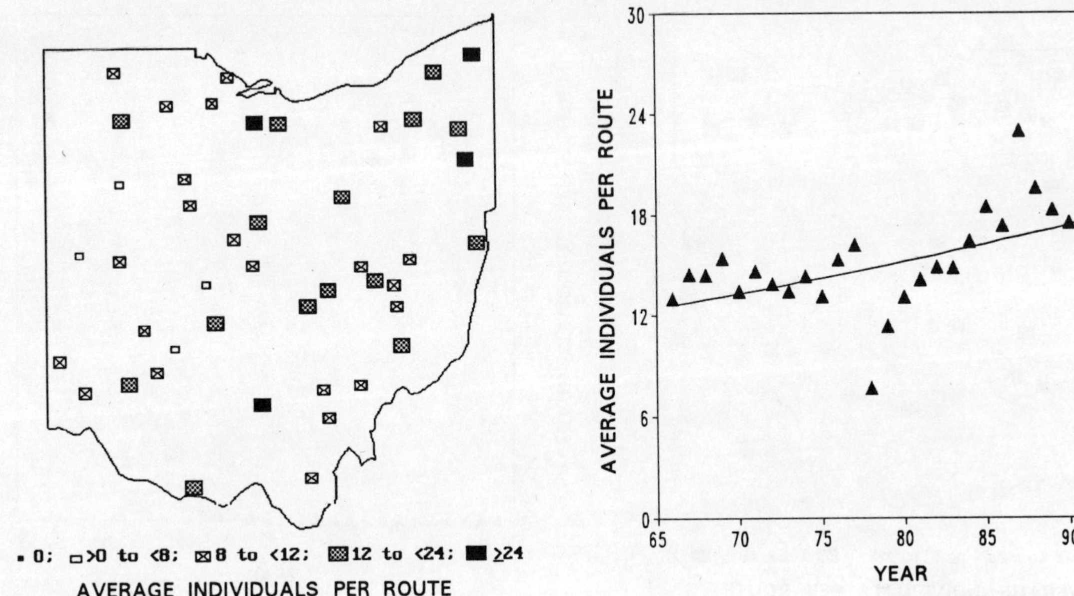

MIGRATORY STATUS: Permanent resident or temperate migrant. There is substantial migratory movement, but some individuals winter in Ohio.

BREEDING HABITAT: Shrub, including brushy woodland edges and abandoned fields.

ABUNDANCE AND DISTRIBUTION: Abundant (15.2 birds per route) and widely distributed (all routes). Abundance in Western and Eastern Ohio did not differ significantly (14.0 vs. 16.6 birds per route, P = 0.09).

OHIO POPULATION TREND: Percent Annual Change (± SE) = 1.3 (± 0.6), P = 0.03
Overall, American Goldfinches have increased significantly at 1.3% annually. The population was fairly stable through 1977 (0.7% annual change, P = 0.51), declined sharply after the severe winter of 1977-78, and has increased significantly at 3.7% annually since 1978 (P < 0.001).
Goldfinches increased significantly in Western Ohio (2.6%, P = 0.001) but exhibited large variation among routes in Eastern Ohio (0.3%, P = 0.74). The difference between trends was significant (P = 0.04).

CONTINENTAL AND GREAT LAKES REGIONAL TREND: In contrast, the continental population decreased significantly at 1.0% annually, and the Great Lakes population exhibited no significant trend.

HOUSE SPARROW

Passer domesticus

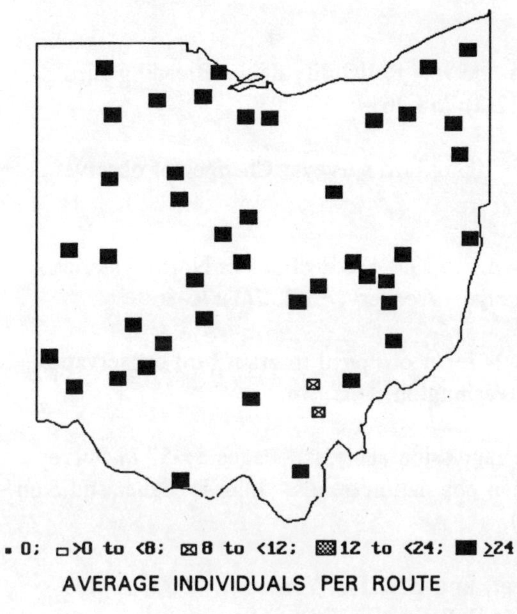

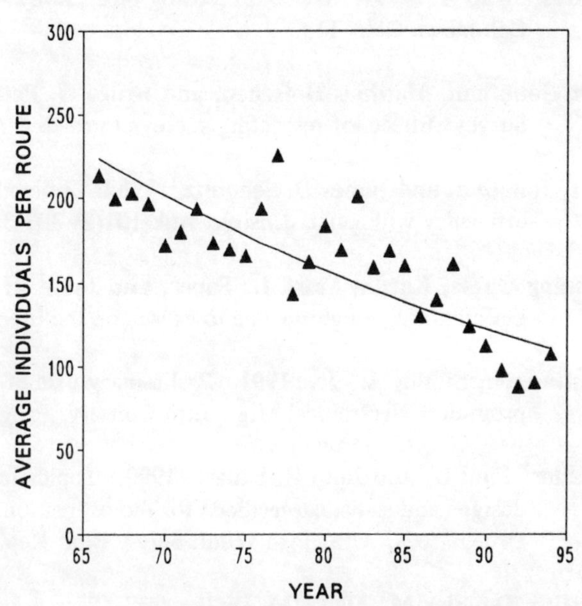

MIGRATORY STATUS: Permanent resident.

BREEDING HABITAT: Residential areas.

ABUNDANCE AND DISTRIBUTION: Abundant (161 birds per route) and widely distributed (all routes). House Sparrows were >3 times as abundant in Western than Eastern Ohio (235 vs. 77 birds per route, P < 0.001).

OHIO POPULATION TREND: Percent Annual Change (± SE) = -2.5 (± 0.4), P < 0.001
  House Sparrows decreased significantly at 2.5% annually. They decreased significantly in Western Ohio at 2.9% annually (P < 0.001) but not in Eastern Ohio (-1.2%, P = 0.19).

CONTINENTAL AND GREAT LAKES REGIONAL TREND: Similarly, the regional and continental population decreased significantly at 1.4 and 1.9% annually.

LITERATURE CITED

Andres, Brad A. 1989. Use of the Breeding Bird Survey to monitor Ohio's nongame bird populations. Unpubl. report, Ohio State University, Columbus, OH. 119p.

Andres, Brad A. 1991. Wetland breeding bird survey - 1991. Unpubl. report, Ohio State University, Columbus, OH. 11p.

Bart, Jonathan, Matthias Hofschen, and Bruce G. Peterjohn. 1995. Reliability of the Breeding Bird Survey: Effects of restricting surveys to roads. Auk 112(3):758-761.

Bart, Jonathan and James D. Schoultz. 1984. Reliability of singing bird surveys: Changes of observer efficiency with avian density. Auk 101(2):307-318.

Bohning-Gaese, Katrin, Mark L. Taper, and James H. Brown. 1993. Are declines in North American insectivorous songbirds due to causes on the breeding range? Conserv. Biol. 7(1):76-86.

Gauthreaux, Sidney A., Jr. 1991. Preliminary lists of migrants for neotropical migrant bird conservation program. Neotropical Migr. Bird Conserv. Program, Washington, D.C. 3p.

Geissler, Paul H. and John R. Sauer. 1990. Topics in route regression analysis. Pages 54-57 *in* Survey designs and statistical methods for the estimation of avian population trends. John R. Sauer and Sam Droege, eds. U.S. Fish Wildl. Serv., Biol. Rep. 90(1).

Griffith, Douglas M., Dawn M. DiGiovanni, Teresa L. Witzel, and Eric H. Wharton. 1993. Forest statistics for Ohio, 1991. U.S. Forest Service, Resource Bulletin NE-128. 169p.

Hagan, John M., III, Trevor L. Lloyd-Evans, Jonathan L. Atwood, and D. Scott Wood. 1992. Long-term changes in migratory landbirds in the northeastern United States: evidence from migration capture data. Pages 115-130 *in* Ecology and conservation of Neotropical migrant landbirds. John M. Hagan III and David W. Johnsnton, eds. Smithsonian Inst. Press, Washington, D.C. 609p.

James, Frances C., David A. Wiedenfeld, and Charles E. McCulloch. 1992. Trends in breeding populations of warblers: declines in the southern highlands and increases in the lowlands. Pages 43-56 *in* Ecology and conservation of Neotropical migrant landbirds. John M. Hagan III and David W. Johnsnton, eds. Smithsonian Inst. Press, Washington, D.C. 609p.

Knopf, Fritz L. 1994. Avian assemblages on altered grasslands. Studies in Avian Biology No. 15:247-257.

Link, W.A. and J.R. Sauer. 1994. Estimating equation estimates of trends. Bird Populations 2:23-32.

National Geographic Society. 1987. Field guide to the birds of North America. Second ed. National Geographic Society, Washington, D.C. 464p.

Peterjohn, Bruce G. 1989. The birds of Ohio. Indiana Univ. Press, Bloomington, IN. 237p.

Peterjohn, Bruce G., R.L. Hannikman, J.M. Hoffman, and E.J. Tramer. 1987. Abundance and distribution of the birds of Ohio. Ohio Biol. Surv. Biol. Notes No. 19. 52p.

Peterjohn, Bruce G. and Daniel L. Rice. 1991. The Ohio Breeding Bird Atlas. Ohio Department of Natural Resources, Division of Natural Areas and Preserves. Columbus, OH. 416p.

Peterjohn, Bruce G. and John R. Sauer. **1994.** Population trends of woodland birds from the North American Breeding Bird Survey. Wildl. Soc. Bull. 22:155-161.

Peterjohn, Bruce G., John R. Sauer, and Chandler S. Robbins. **1996.** Population trends from the North American Breeding Bird Survey. Pages 13-27 *in* Ecology and management of Neotropical migratory birds: a synthesis and review of the critical issues. Thomas E. Martin and Debra M. Finch, eds. Oxford University Press, Oxford, UK.

Rappole, John H., Eugene S. Morton, Thomas E. Lovejoy, James L. Ruos. **1983.** Nearctic avian migrants in the Neotropics. U.S. Fish & Wildl. Serv., U.S. Dept. of Interior. 646p.

Robbins, Chandler S., Danny Bystrak, and Paul H. Geissler. **1986.** The Breeding Bird Survey: its first fifteen years, 1965-1979. U.S. Fish & Wildl. Serv. Resour. Publ. 157. 196p.

Robbins, Chandler S., John R. Sauer, Russell S. Greenberg, and Sam Droege. **1989.** Population declines in North American birds that migrate to the neotropics. Proc. Natl. Acad. Sci. 86:7658-7662.

Sauer, John R. and James B. Bortner. **1990.** Population trends from the American Woodcock singing-ground survey, 1970-88. J. Wildl. Manage. 55(2):300-312.

Sauer, John R. and Sam Droege. **1992.** Geographic patterns in population trends of Neotropical migrants in North America. Pages 26-42 *in* Ecology and conservation of Neotropical migrant landbirds. John M. Hagan III and David W. Johnston, eds. Smithsonian Inst. Press, Washington, D.C. 609p.

Sauer, John R., Bruce G. Peterjohn, and William A. Link. **1994.** Observer differences in the North American Breeding Bird Survey. Auk 111(1):50-62.

Steele, Robert G.D. and James H. Torrie. **1980.** Principles and procedures of statistics: a biometrical approach. Second Edition. McGraw-Hill Book Company, New York, NY. 633p.

Thomas, Len. **1996.** Monitoring long-term population change: why are there so many analysis methods? Ecology 77(1):49-58.

Witham, Jack W. and Malcolm L. Hunter, Jr. **1992.** Population trends of Neotropical migrant landbirds in northern coastal New England. Pages 85-95 *in* Ecology and conservation of Neotropical migrant landbirds. John M. Hagan III and David W. Johnsnton, eds. Smithsonian Inst. Press, Washington, D.C. 609p.

APPENDIX I

STATISTICAL ANALYSIS

by Jonathan Bart and Susan L. Earnst

Several different methods have been proposed for estimating temporal trends in abundance from count data (e.g., Thomas 1996). These methods differ in the definition of "trend", in assumptions about whether observer ability varies, and in the specific analytic methods used to estimate trend. In this appendix, we explain the method we used, relate it to other methods, and compare the results we obtained with results obtained using two other methods. The methods used in this report will be described in more detail elsewhere (Bart and Notz, in prep.)

Trend estimation

An exponential curve was fitted to the mean number of birds recorded per route in each year. Curve fitting was accomplished by taking logs of each annual mean and analyzing these with linear regression. The trend was then taken as the exponent of the resulting formula (i.e., value of the trend in year t $= e^{\alpha + \beta t}$ where $\alpha =$ intercept and $\beta =$ slope from the linear regression). Note that this approach does not require an assumption that the trend on <u>each route</u> follows an exponential curve.

Statistical issues arise because the counts are not independent (the same routes are sampled each year). The estimator of the trend described above is unbiased but new formulas for the variance of this estimate were needed. To obtain them, the estimated regression coefficient, b, was expanded in a Taylor Series about the true annual means. The variance of b can then be estimated as

$$v(b) \approx \sum_{j1}^{L} \sum_{j2}^{L} \left(\frac{C_{j1} C_{j2}}{\overline{y}_{j1} \overline{y}_{j2}} \right) cov(\overline{y}_{j1}, \overline{y}_{j2})$$

where L is the number of years (29), C_{j1} and C_{j2} are known constants (functions of the years), $\overline{y}_{j1}$ and $\overline{y}_{j2}$ are the sample means from year j1 and j2, and $cov(\overline{y}_{j1}, \overline{y}_{j2})$ is the sample covariance of $\overline{y}_{j1}$ and $\overline{y}_{j2}$. We verified that this formula provided essentially unbiased estimates (i.e., bias $< 5\%$) by use of a simulation program. The simulation also showed that the test statistic (used in hypothesis testing) had a t distribution and thus that t tables could be used in the analysis. The degrees of freedom were n - 2 where n was the number of years.

Geissler and Sauer (1990) point out that estimating trends by carrying out linear regression on the logs of the number recorded (the approach we used) produces a biased estimate of the exponential trend. Although Bradu and Mundlak (1970) provide a method for removing this bias, it requires that the observations be independent which is true for counts on individual routes, but not for annual means. A simulation showed that the bias in our estimates is extremely small ($< 5\%$).

Comparison with route regression

The route regression method for analyzing BBS data, used by researchers at the BBS Office, involves fitting an exponential curve to each route and then taking the overall trend as the mean of the route-specific trends (e.g., Geissler and Sauer 1990, Sauer et al. 1994). The route regression method includes provisions for weighting results from each route to account for differences in abundance between routes, and replacing zero-counts with an arbitrary constant before logs of counts are taken and exponential regression is performed. A recent approach using estimating equations avoids the need to add a constant to zero counts (Link and Sauer 1994).

It is important to realize that regression on the annual means (our method) and route regression are estimating different quantitites and thus sometimes produce different numbers as "the trend". Suppose, for example, that we had only five years of data and five routes, that a single observer collected all the data, and that the results were as shown in Table 1. Trends along individual routes vary, and no clear trend exists in the means. Route regression (with equal weighting and the bias correction described by Geissler and Sauer 1990) produces a trend of 0.982, or an annual decrease of about 2%. Regression on the means produces a trend of 1.000, or 0% annual change. Two points can be made from this example, and looking at hundreds of graphs of Ohio BBS data analyzed using route regression and regression on the means (as we have done) reinforces them. First, route regression and regression on the means are estimating different parameters. In many cases the results are similar using both methods, but we should not expect them to be identical. Second, route regression produces estimates of the trend that do not always fit the means very well. Thus, in the example above, the exponential curve fit by least squares methods has an annual change of 1.0 (i.e., no change) whereas the route regression line has an annual change of 0.982. We developed an approach using regression on the means because we wanted the trend lines to fit the annual means (presumed to reflect annual population size) as closely as possible (i.e., according to least-square curve fitting procedures) and route regression often does not. This realization, however, does not make one method correct and the other incorrect; the two methods and the parameters that they estimate are simply different.

Table 1. Hypothetical counts on five survey routes during five years.

| | Year | | | | |
Route	1	2	3	4	5
1	18	12	9	15	14
2	11	13	20	18	16
3	26	29	24	27	18
4	10	10	9	13	15
5	13	15	20	17	10
Average	15.6	15.8	16.4	18.0	14.6

Change in observer ability

Sauer et al. (1994) have shown that in some BBS data sets average observer skill improves through time causing changes in the mean number of birds reported even if the population is actually stable. It is common for a route to be run by more than one sequential observer, thus creating the potential for observer differences to influence the trend estimate. In Ohio, most routes (78%) had ≤ 4 different observers and many routes (25%) had only 1 or 2 different observers.

We incorporated observer effects by performing route regression on the number of birds seen in each year and coding observers as indicator variables. The regression coefficients of the observer variables were then used to correct the original data before calculating annual means and performing exponential regression on those means. The method used by the BBS Office obtains observer coefficients in the same way, however, it then estimates the overall trend as the average of the route trends (see above).

We investigated two models of observer effects. The "full observer effect" model included all observer effects (similar to Sauer et al. 1994). The "partial observer effect" model included only observer effects that explained a significant proportion of the variance in number of birds per route (similar to Beavers and Ramsey, in press). The "full effects" model requires estimation of 153 variables (for each route, it estimates slope, intercept, and k observer variables, where k = number of observers minus 1). As others (e.g., James et al., 1996) have noted, estimating this large number of parameters reduces precision. Also, we found that unreasonably large observer coefficients are sometimes generated on routes or for species with relatively poor data. In particular, unreliable estimates appear to arise on routes that have one or more observers that participate for only a few years, have few birds recorded per year for the species of interest, or have an underlying trend that is not exponential.

The "partial observer effect" model was more satisfactory than the "full" model simply because it estimated fewer coefficients and also because some of the more unreliable coefficients occurred on routes for which the set of observer variables did not explain a significant proportion of the variance and were thus excluded. To further reduce the number of extreme coefficients in the "partial effects" model, we used coefficients only if they fell in a moderate range, otherwise we used the endpoints of the range. Visual inspection of numerous routes indicated that coefficients outside this range were usually caused by non-exponential trends or poor data (see above) rather than by extreme differences in observer ability. The range that we used, 0.35 to 2.9, encompassed 54% of observer effects. Note that an observer effect of 0.35 indicates that data from the observer in question divided by 0.35 would be comparable to data from Observer 1.

Comparison of results among methods

Using the "partial observer effect" model, we calculated the 29-year population trend for each of the 105 species in the Species Accounts and compared the results to our original analysis, which did not include observer effects, and to trends provided by the BBS Office (Peterjohn, pers. comm.). The "partial" model gave very similar results to our original model. The average difference in percent annual change among all species was only 0.04%. Thus, we chose to use the original model because it will be more easily interpreted by the readers of this monograph and because it estimates fewer parameters.

The average difference in percent annual change using the BBS Office model vs. the original was 0.95; the BBS Office model tended to produce trends that were more negative. Nonetheless, our original trend estimates and those of the BBS Office are highly correlated ($r^2 = 0.89$), and there were no species for which the methods produced significant trends in opposite directions (i.e., one increasing and one decreasing). Most differences in conclusions were due to one method's estimate being statistically significant and the other's being in the same direction but not significant.

Because our most important conclusion is the comparison of proportion of declining species among habitat types, we repeated the comparison using trends calculated with the partial observer effect model and trends supplied by the BBS Office. Using the original method, we concluded that species breeding in grasslands are declining proportionately more than any other group and that there is no evidence to suggest that mature forest species are incurring proportionately more declines than scrub or nonspecific forest species. The same conclusions would be reach using either the partial observer effect model or the trends supplied by the BBS Office (Table 2).

Table 2. Percent of species exhibiting significantly decreasing trends under our original model (no observer effects), a partial observer effect model, and the method used by the BBS Office.

Breeding Habitat	Original	Partial Observer Effect	BBS Office
Grasslands	71	71	67
Nonspecific Woods[a]	19	15	19
Scrub/Young Woods	11	11	13
Mature Woods	10	10	19

[a]Includes species occurring in both mature and scrub/young woods (i.e., not easily assigned to one category or the other), and species occurring in a forest-grassland mosaic (e.g., Red-tailed Hawk and Orchard Oriole).

Literature Cited

Beaver, Sallie C. and Fred L. Ramsay. **In press.** Detectability analysis in transect surveys. J. Wildl. Manage.

Bradu, Dan and Yair Mundlak. **1970.** Estimation in lognormal linear models. J. Amer. Statistical Assoc. 65:198-211.

Geissler, Paul H. and John R. Sauer. **1990.** Topics in route regression analysis. Pages 54-57 in J.S. Sauer and S. Droege, eds., Survey designs and statistical methods for the estimation of avian population trends. U.S. Fish and Wildlife Service, Biological Report 90(1).

James, Frances C., Charles E. McCulloch, and David A. Wiedenfeld. **1996.** New approaches to the analysis of population trends in land birds. Ecology 77(1):13-27.

Link, W.A. and J.R. Sauer. **1994.** Estimating equation estimates of trends. Bird Populations 2:23-32.

Sauer, John R., Bruce G. Peterjohn, and William A. Link. **1994.** Observer differences in the North American Breeding Bird Survey. Auk 111(1):50-62.

Thomas, Len. **1996.** Monitoring long-term population change: why are there so many analysis methods? Ecology 77(1):49-58.

APPENDIX II

Species rarely observed breeding during the Ohio Atlas Project

For the purpose of this publication, rare breeders are defined as those that were confirmed or probable breeders on <15 Atlas blocks or special areas during the Ohio Atlas Project.

Double-crested Cormorant (*Phalacrocorax auritus*)
American Bittern (*Botaurus lentiginosus*)
Great Egret (*Casmerodius albus*)
Snowy Egret (*Egretta thula*)
Little Blue Heron (*Egretta caerulea*)
Cattle Egret (*Bubulcus ibis*)
Black-crowned Night-Heron (*Nycticorax nycticorax*)
Yellow-crowned Night-Heron (*Nycticorax violaceus*)
Mute Swan (*Cygnus olor*)
Green-winged Teal (*Anas crecca*)
Northern Pintail (*Anas acuta*)
Northern Shoveler (*Anas clypeata*)
Gadwall (*Anas strepera*)
American Wigeon (*Anas americana*)
Redhead (*Aythya americana*)
Ruddy Duck (*Oxyura jamaicensis*)
Black Vulture (*Coragyps atratus*)
Bald Eagle (*Haliaeetus leucocephalus*)
Northern Harrier (*Circus cyaneus*)
King Rail (*Rallus elegans*)
Sandhill Crane (*Grus canadensis*)
Laughing Gull (*Larus atricilla*)
Ring-billed Gull (*Larus delawarensis*)
Common Tern (*Sterna hirundo*)

Black Tern (*Chlidonias niger*)
Barn Owl (*Tyto alba*)
Long-eared Owl (*Asio otus*)
Short-eared Owl (*Asio flammeus*)
Northern Saw-whet Owl (*Aegolius acadicus*)
Chuck-will's-widow (*Caprimulgus carolinensis*)
Yellow-bellied Sapsucker (*Sphyrapicus varius*)
Red-breasted Nuthatch (*Sitta canadensis*)
Bewick's Wren (*Thryomanes bewickii*)
Winter Wren (*Troglodytes troglodytes*)
Hermit Thrush (*Catharus guttatus*)
Loggerhead Shrike (*Lanius ludovicianus*)
Bell's Vireo (*Vireo bellii*)
Golden-winged Warbler (*Vermivora chrysoptera*)
Magnolia Warbler (*Dendroica mognolia*)
Blackburnian Warbler (*Dendroica fusca*)
Northern Waterthrush (*Seiurus noveboracensis*)
Mourning Warbler (*Oporornis philadelphia*)
Canada Warbler (*Wilsonia canadensis*)
Lark Sparrow (*Chondestes grammacus*)
Dark-eyed Junco (*Junco hyemalis*)
Yellow-headed Blackbird (*Xanthocephalus xanthocephalus*)
Pine Siskin (*Carduelis pinus*)

INDEX TO COMMON NAMES

PUBLICATION POLICY OF THE OHIO BIOLOGICAL SURVEY

Any manuscript to be considered by the Ohio Biological Survey must first and foremost be scientifically credible and accurate. The Survey is most interested in manuscripts which concentrate on Ohio's biota. A monograph outlining the status of a particular taxon within Ohio, an in-depth study of a particular Ohio natural area, or an amalgamation of information of use to Ohio's biologists are examples of areas of paramount interest to the Survey.

Secondly, the Ohio Biological Survey will consider a manuscript dealing with a topic in which Ohio, or an Ohio natural area, or an Ohio native taxon is an integral part. Any study encompassing the Ohio River drainage basin, the Great Lakes region, or eastern North America, for example, would be given a reading by the Ohio Biological Survey.

Finally, the Ohio Biological Survey will consider manuscripts by Ohio authors. It is preferable that these manuscripts meet either of the criteria above, and such manuscripts will be given priority. However, if an Ohio author creates a manuscript meeting the criteria for one of the Survey's publication categories, it will be read with an eye to publication.

It should be remembered, however, that any manuscript, regardless of origin or topic, must be first approved by the Survey's Editorial Committee, and then pass peer-review by several experts in the field of study.

Ohio Biological Survey
1315 Kinnear Road
Columbus, OH 43212-1192

PUBLICATIONS CRITERIA

I. Ohio Biological Survey *Bulletin New Series*

1. Original work
2. Data-rich, and therefore probably lengthy
3. Comprehensive
4. Having potential archival value
5. Historical perspective on the problem plus the rationale for the current investigation
6. "Literature Cited" must be included
7. If a taxonomic work, synonymies and diagnoses should be included and new taxa may be raised; checklists, distribution data, and keys with accompanying figures should be included
8. Bibliography, Foreword, Preface, Ecological/Behavioral Information—these sections are optional
9. Manuscript will be peer-reviewed by two or more experts in the field

II. Ohio Biological Survey *Miscellaneous Contribution*

1. Original work
2. Of variable length but usually shorter than a *Bulletin*
3. The category *Miscellaneous Contributions* may encompass proceedings of meetings, annotated bibliographies, anthologies, catalogs, atlases, and others
4. Where appropriate, *Miscellaneous Contributions* will have a "Literature Cited" section, or perhaps more than one such section
5. Manuscripts will be peer-reviewed by two or more experts in the field

III. Ohio Biological Survey *Notes*

1. Original work
2. Focused on a precisely defined topic
3. Of variable length, but concise organization
4. *Notes* will present the results of biological surveys, unannotated bibliographic compilations, unified short biological studies, and/or short natural historical offerings; several short papers of a theme may be combined
5. Where appropriate, *Notes* will have a "Literature Cited" section, or perhaps more than one such section
6. Manuscript will be peer-reviewed by two or more experts in the field

IV. Ohio Biological Survey *Informative Publications*

1. Quickly-produced and disseminated amalgamations of information useful (primarily) to Ohio biologists (lists; bibliographies; meetings' abstracts)
2. Short; likely to be the last word on a subject for only a brief time
3. Internal review only

REQUIREMENTS FOR AUTHORS

Please refer first to the Publications Criteria listed previously.

I.   Sections, and suggested order (please bear in mind that not all sections need be included, and certain sections are neither necessary or even desirable for all four publication categories).

> A.   Abstract
> B.   Acknowledgements
> C.   Lists of (Figures, Tables, Maps, Photos, Plates)
> D.   Introductory Materials (could include behavior, ecology, geology, history—these could be subsections)
> E   Materials and Methods
> F.   Checklist
> G.   Results (could be set up as subsections)
> H.   Discussion (again, would probably be set up in subsections, especially for the *Bulletins New Series*, *Miscellaneous Contributions*, and *Notes*, and could include data analyses, diagnoses, population information, new taxa and the rationale for their creation)
> I.   Taxonomic Key(s)
> J.   Summary
> K.   Glossary
> L.   Literature Cited
> M.   Bibliography
> N.   Appendices
> O.   Indices

Prose style should be that of previously published *Bulletins New Series*, *Miscellaneous Contributions*, *Biological Notes*, or *Informative Circulars*. Spell out numbers one through ten, and use Arabic numerals for 11 and above. Use the active voice. Scientific names and the titles of books and journals will be italicized in the published work (see II. below). Abbreviations and symbols used should be used following the guidelines and lists given in the CBE Style Manual, 5th Edition. Use the International System of Units in all cases.

NEVER use footnotes.

II.   Manuscripts should be submitted on computer disk as a Word Perfect for Windows file (the most recent version is preferred). The format of earlier Ohio Biological Survey publications should be followed. Scientific names and the titles of books and journals should be italicized. Tables should be placed in Word Perfect table format. If Pagemaker or QuarkExpress for Windows is used, then black and white artwork can be scanned right into place, as can black and white photographs. However, the original graphics and photographs must accompany the manuscript file.

III.   Black and white photographs should be submitted as prints. Black and white line drawings should be submitted as PMTs (PMT is an abbreviation of "photomechanical transfer," and refers to the process by which a black and white piece of artwork is reduced or enlarged photographically, and then transferred mechanically in between two pieces of receiver paper and treated with an activator fluid, so that you end up with a positive representation of the drawing you want at the size you want. They are sometimes called "stats," as well). Color photographs should be submitted as transparencies — slides or photographs up to 8.5" x 11."